2020 AISYIYAH International Conference on Engineering Science and Its Applications (A-ESIA 2020)

IOP Conference Series: Materials Science and Engineering
Volume 1232

Yogyakarta, Indonesia
30 June – 1 July 2020

Editors:

Deborah Davis
Linda McGowan
Ismail bin Said
Taufiqur Rahman
Wantonoro
Mufdlilah
Besral
Ramadiani
Sunardi
Tri Hastuti Nur Rochimah
Andri Pranolo
Andari Wuri Astuti
Luluk Rosida

ISBN: 978-1-7138-4690-1
ISSN: 1757-8981

Printed from e-media with permission by:

Curran Associates, Inc.
57 Morehouse Lane
Red Hook, NY 12571

Some format issues inherent in the e-media version may also appear in this print version.

Institute of Physics
Dirac House, Temple Back
Bristol BS1 6BE UK

Phone: 44 1 17 929 7481
Fax: 44 1 17 920 0979

techtracking@iop.org

Additional copies of this publication are available from:

Curran Associates, Inc.
57 Morehouse Lane
Red Hook, NY 12571 USA
Phone: 845-758-0400
Fax: 845-758-2633
Email: curran@proceedings.com
Web: www.proceedings.com

TABLE OF CONTENTS

IOP Conf. Series: Materials Science and Engineering 1232 (2022) 011001 doi:10.1088/1757-899X/1232/1/011001

PREFACE

People's quality of life recently becomes an interesting-debates across professionals. World Health Organisation (WHO) stated that improving the people's health and well-being is an important-factors contribute to people's quality life. Their health and well-being can help the mengaging 4.0 industrial revolution era and it also determines the health of the next generation and can help predict future of public health challenges for families, communities, and the health care system. Therefore, research, policies and strategies have been conducted and promoted in order to achieve the goal in each national level including Indonesia through innovation and involvement both public and private sectors. Additionally, the rapid development of science and technology in the 21st Century offers more benefits and opportunities. At the same time, it adds more challenges and threats to society. The development of technology has enabled people to live longer and more comfortably, yet at the same time have created a destructive lifestyle for both natural resources and socialties. Lifestyle in the Digital Age demands all things fast paced, instant and sophisticated, and easy, without the need for physical activity and tendstobestatic. To address such complex situations, stake holder collaboration is needed between the Government, the private sector, civil society and scholars. It also needs to be highlighted that research activity is crucial for improving health and well-being because it enables academicians and practitioners to explore issues and provide evidence for those who concerning for in national or international level. Additionally, the one that also crucial to be focused on, is a research dissemination which enables researcher, professionals, health care providers, students or evenly person disseminate, share and discuss about topic matters.

2020 AISYIYAH International Conference on Engineering Science and Its Applications (A-ESIA) as part of 2nd International Conference of Health, Science and Technology which we originally planned to be placed in June 2020 by using conventional face to face meeting, however, due to Covid-19 impact it therefore changed to be run in Juli 2020 by using online platform. It was because the committee were considering about travel restriction and we also put people's safety first during Covid-19 pandemic situation. The virtual Conference held on 18th July 2020 at 07.00 AM -17.30 PM in Jogjakarta Location of the conference at Aisyiyah University Street West Ringroad No.63, Nogotirto, Gamping, Sleman Regency, Yogyakarta Special Region 55592via zoom meeting conference and consists of 5 plenary sessions.

The virtual conference was run successfully with enthusiasm from all participants. There were 96 speakers with various topical and strategic papers participated into this 2nd International Conference of Health, Science and Technology, this means that the number of participants has increased from previous year. We would like to thank to all participants, committee and publication team whom contribute to the2nd International Conference of Health, Science and Technology. We do also really appreciate fo all reviewers whom made valuable suggestions for the authors to improve their works. Also, special thanks go to IOP conference series for making the publication possible.

We hope that this conference is strengthen a great academic atmosphere to the academicians in the fields of health, as well as promote scientific publication. We hope that the publications can make evidence and information to the broader society and eventually giving benefits for today's development. Finally, on be-half of the committees, we would like to thank your valuable participants for the active and fruitful contribution to this conference. We look forward to seeing all of you next year at the conference.

Sarwinanti, M.Kep.,Sp Mat
General Chair of 2020 AISYIYAH International Conference on Engineering Science and Its Applications (A-ESIA. Part of 2nd International Conference on Health Science and Technology (ICHST) 2020

IOP Conf. Series: Materials Science and Engineering 1232 (2022) 011001 doi:10.1088/1757-899X/1232/1/011001

EDITORS A-ESIA

Editor	Organization	Email
Prof. Deborah Davis	University of Canberra, Australia	Deborah.Davis@canberra.edu.au
Prof. Linda Mc Gowan	University of Leeds, United Kingdom	L.McGowan@leeds.ac.uk
Prof. Dr. Ismail bin Said	Universiti Teknologi Malaysia, Malaysia	ismailsaaid@utp.edu.my
Taufiqur Rahman, Ph.D.	Universitas Muhammadiyah Yogyakarta, Indonesia	taufiqurrahman@umy.ac.id
Wantonoro, Ph.D.	Universitas Aisyiyah Yogyakarta, Indonesia	oneto_ns@yahoo.com
Dr. Mufdlilah, M.Sc.	Universitas Aisyiyah Yogyakarta, Indonesia	oneto_ns@yahoo.com
Dr. Besral, S.Km., M.Sc.	Universitas Indonesia	besral@ui.ac.id
Ramadiani, M.Kom., Ph.D.	Universitas Mulawarman, Indonesia	ramadiani@fkti.unmul.ac.id
Sunardi, Ph.D.	Universitas Ahmad Dahlan, Indonesia	sunardi@mti.uad.ac.id
Dr. Tri Hastuti Nur Rochimah, M.Si.	Universitas Muhammadiyah Yogyakarta, Indonesia	trinur@umy.ac.id
Andri Pranolo	Universitas Ahmad Dahlan, Indonesia	andri.pranolo@tif.uad.ac.id
Andari Wuri Astuti, Ph.D.	Universitas Aisyiyah Yogyakarta, Indonesia	astutiandari@unisayogya.ac.id
Luluk Rosida, S.ST., M.KM	Universitas Aisyiyah Yogyakarta, Indonesia	lulurosida@unisayogya.ac.id

Committee	Organizer	Organization
Ns Sarwinanti, Mkep., Sp.Kep.Mat.	General Chair	Universitas Aisyiyah Yogyakarta, Indonesia
Enny Fitriahadi, S.Si.T. M.Kes.	Co Chair	Universitas Aisyiyah Yogyakarta, Indonesia

IOP Conf. Series: Materials Science and Engineering 1232 (2022) 011001 doi:10.1088/1757-899X/1232/1/011001

Prof. Deborah Davis	Scientific Committee	University of Canberra, Australia
Dr. Gherissi Atf	Scientific Committee	University of Tunis El Manar, Tunisia
Prof. Linda Mc Gowan	Scientific Committee	University of Leeds, United Kingdom
Prof. Dr. Ismail bin Said	Scientific Committee	Universiti Teknologi Malaysia, Malaysia
Taufiqur Rahman, Ph.D.	Scientific Committee	Universitas Muhammadiyah Yogyakarta, Indonesia
Wantonoro, Ph.D.	Scientific Committee	Universitas Aisyiyah Yogyakarta, Indonesia
Dr. Mufdlilah, M.Sc.	Scientific Committee	Universitas Aisyiyah Yogyakarta, Indonesia
Dr. Besral, S.Km., M.Sc.	Scientific Committee	Universitas Indonesia
Ramadiani, M.Kom., Ph.D.	Scientific Committee	Universitas Mulawarman, Indonesia
Sunardi, Ph.D.	Scientific Committee	Universitas Ahmad Dahlan, Indonesia
Dr. Tri Hastuti Nur Rochimah, M.Si.	Scientific Committee	Universitas Muhammadiyah Yogyakarta, Indonesia
Andri Pranolo, M.Cs.	Scientific Committee	Universitas Ahmad Dahlan, Indonesia
Andari Wuri Astuti, Ph.D.	Scientific Committee	Universitas Aisyiyah Yogyakarta, Indonesia
Luluk Rosida, S.S.T., M.K.M.	Organizing Committee	Universitas Aisyiyah Yogyakarta, Indonesia
Erwin Rasyid, S.I.Kom., M.Sc.	Organizing Committee	Universitas Aisyiyah Yogyakarta, Indonesia
Ellyda Rizki Wijhati, S.ST., M.Keb.	Organizing Committee	Universitas Aisyiyah Yogyakarta, Indonesia
Sri Sugesti	Organizing Committee	Universitas Aisyiyah Yogyakarta, Indonesia

Peer review declaration

All papers published in this volume of IOP Conference Series Materials Science and Engineering Journal have been peer-reviewed through processes administered by the Editors. Expert referees reviewed the professional and scientific standards expected of a proceedings journal published by IOP Publishing.

- **Type of peer review: Single-blind**

  Each submitted paper reviewed by two minimum of reviewers after meet the minimum criteria. The review based on the following aspects: 1) Technical Criteria (Scientific merit, Clarity of expression, and Sufficient discussion of the context of the work, and suitable referencing); 2) Quality Criteria (Originality, Motivation, Repetition, Length); and 3) presentation criteria (Title, Abstract, Diagram, figures, tables and captions, Text and mathematics, and Conclusion). We also used iThenticate for plagiarism detection.

  - **Conference submission management system: Easychair**
  - **Number of submissions received: 92**
  - **Number of submissions sent for review: 92**
  - **Number of submissions accepted: 82 (14 submitted for IOP MSE Scope, Other Submitted to Health Science Series of Atlantis Press)**
  - **Acceptance Rate (Number of Submissions Accepted / Number of Submissions Received X 100): 89%**
  - **The average number of reviews per paper: 2**
  - **Total number of reviewers involved: 19**
  - **Any additional info on the review process: No**
  - **Contact person for queries:**

Contact Person:
Sri Sugesti (+62877-3980-3053)
Sarwinanti (+62815-7848-2820)

Address:
Kampus Universitas 'Aisyiyah Yogyakarta
Jalan Siliwangi (Ring Road Barat) No 63, Mlangi, Nogotirto, Gamping, Sleman, Yogyakarta, Indonesia

E-mail:
ichst@unisayogya.ac.id

IOP Conf. Series: Materials Science and Engineering 1232 (2022) 011002 doi:10.1088/1757-899X/1232/1/011002

Website:
https://ichst.unisayogya.ac.id/2020

NiO Supported on Porous Carbon From Merbau Wood: Synthesis and Characterization

D Santi[1,2], Triyono[2], W Trisunaryanti[2] and I I Falah[2]

[1] Department of Chemistry, Faculty of Mathematics and Natural Sciences, Universitas Papua, Manokwari, Indonesia.
[2] Department of Chemistry, Faculty of Mathematics and Natural Sciences, Universitas Gadjah Mada, Yogyakarta, Indonesia.

d.santi@unipa.ac.id, triyn102@ugm.ac.id, wegats@ugm.ac.id

Abstract. Nickel oxide catalyst supported on porous carbon (NiO/C), the materials support were synthesized from Merbau wood. The metals were loaded onto the carbon by the wet impregnation method. After drying and calcination, NiO/C was obtained. The prepared catalyst was evaluated for the hydrocracking of pyrolyzed α-cellulose. The catalyst was characterized by XRD, SEM-EDX, FT IR, SAA, and acidity test by NH_3 base vapor adsorption determined gravimetrically and by FT IR. The XRD results showed that the carbonaceous of materials support were amorphous materials that characteristic for mesopore carbon. The impregnation of nickel into carbon results in difraktogram peaks that correspond to the presence of NiO, which was confirmed by EDX. The SEM image showed that the morphology of the catalyst samples was a honeycomb-like structure. The acidity test showed that the total acid amount of NiO/C samples was 3.05 mol/g, and the FT IR spectra indicated that there were Brönsted and Lewis acid sites on NiO/C catalyst. The SAA results showed that the specific surface area, total pore volume, and average pore diameter of the NiO/C catalyst were 520.41 m^2/g, 0.27 cm^3/g, and 2.083 nm, respectively.
Keywords: nickel oxide, Merbau wood, carbon, hydrocracking, α-cellulose

1. Introduction

Activated carbon has been developed as one of the most popular uses, as a material in environmental management technology, used as an electrode material, or as a catalyst support or as a catalyst in catalytic processes over the last decades. It has possessed good surface morphology such as high surface area and high pore volume, pore distribution, and surface functionality for its utilize as catalysts and adsorbents [1–3]. A variety of precursors can prepare to produce activated carbon. Generally, these precursors have volatile and high carbon content but low inorganic contents [4]. It is well known that various carbonaceous precursors can be found in agricultural residues, such as sugarcane bagasse [5], oil-palm fruit waste [6], palm shell and coconut shell [7], cotton stalks [8], pine chips [9], and rice husks [10]. As a renewable, inexpensive, sustainable, and environment-friendly feedstock, biomass has received considerable attention as the potential precursor [11,12].

Furthermore, the high concentration of volatiles is ideal for preparing the activated carbon matrix s within a highly porous structure. The significant constitutive fractions, in the term of weight of these lignocellulose materials as the use as activated carbon precursor, are the hemicellulose, cellulose, and lignin. Merbau wood as lignocellulose materials has been known to have potential as a carbon precursor [13]. Because of the difference in reactivity between these components during the pyrolysis step, the study of their carbonization is rather complicated. The nature of the precursor, conditions, and activation method determined the characteristics of activated carbon porosity properties, including porous domains (microporosity, mesoporosity, and macroporosity), pore size distribution, shapes of the pores, and surface chemistry was determined by Byrne and Marsh [14].

The methods for preparing an activated carbon involves two steps-pyrolysis and activation. In the pyrolysis step, the temperature of carbonization between 400 and 850 °C, sometimes reaches 1000 °C. In the activation proses, these chars as a result of the pyrolysis step activated in oxidizing gas such as air, carbon dioxide, or steam at a temperature of about 800-1000 °C to produce the final activated carbon. The use of high temperature [15] and these types of selected gas result activated carbon with high micropore. Furthermore, the active catalyst support required the large surface area that provides sufficient active sites for loading catalytic metal nanoparticles.

Moreover, mesopore is required for adsorption purposes, diffusion, and accelerated transport [16,17]. The previous study revealed that the lower temperature, 350 °C with oxidized in the air, produced the mesoporous activated carbon [18]. However, there is no systemic study related to the preparation of carbon active derived from Merbau wood (Intsia, spp) and the use of oxygen as oxidizing gas at the temperature of 350 °C via the single-step process.

Cellulose was a representative biomass compound [16], one of the most abundant [17] and an inexpensive feedstock for energy production [18] and conversion into more valuable chemicals [19]. Cellulose conversion into fuels is inspiring and involving effective catalytic systems [18]. It has been reported that transition metals-modified dispersed to porous materials are used as a catalyst in hydrocracking reaction [18–20]. The benefits of this type of catalyst include the inclusion of metal acid sites and the avoidance of metal agglomeration due to the dispersal of transition metals into the porous system such as mesoporous silica [19], MCM-41 [21], and mesoporous carbon [13,18]. In this work, the main study was to investigate the catalytic performance of nickel oxide/carbon catalysts on the hydrocracking of pyrolyzed α-cellulose as a model of biomass.

2. Experimental

2.1. Materials
Merbau wood was obtained from Manokwari, Indonesia. KOH, NaOH, $NaHCO_3$, Na_2CO_3, $Ni(NO_3)_2{\bullet}6H_2O$, acetone, HCl were analytical grade and obtained from Merck, α-cellulose was obtained from Sigma-Aldrich.

2.2. Carbonization and Oxydation of Merbau woods
The Merbau woods were heated up to carbonization temperature of 700 °C at the heating rate 10 °C/min, and were held for 2 h, under N_2 gas at a flow rate of 20 mL/min. The samples were cooled to room temperature under N_2 flow. Activation is carried out by streaming oxygen gas with a flow rate of 15 mL/min for 3 h, 350 °C at a heating rate of 10 °C/min. The results of this process were noted as C.

2.3. Catalysts Preparation
The C samples were soaked for 2 h and then filtered, by acetone and HCl, respectively. This step followed by thorough washing with deionized water and drying at 105 °C for 12 h. The impregnation of nickel (1.0 wt.%) on the C samples was done by the wetness impregnation technique. The solid was then calcined at 500 °C under nitrogen gas stream with a flow rate of 20 mL/min for 4 h. The final products were denoted as NiO/C catalysts. The catalyst was measured acidity value by acidity test using ammonia adsorption method. Characterization of the crystallinity sample was carried out by X-ray diffraction (XRD) using a Rigaku Miniflex600, λ = 1,54 Å, 40 kV, 15 mA. The functional groups as analysis surface chemistry of the samples were recorded between 4000 and 400 cm^{-1} using Fourier Transform Infrared Spectrometer (FTIR, Shimadzu Prestige-21) with KBr disc technique. The

morphology of the sample was characterized by scanning electron microscopy (SEM, Jeoul JSM-6510).

2.4. Catalyst Acidity Test

The acidity value of the catalysts was measured by the absorption of the ammonia vapor onto the NiO/C sample. The acidity value (g/mol) of the catalysts was calculated using the following equation [13] with any modification the type of gas adsorbate using the ammonia vapor:

$$\text{Acidity value} = (G - G_0)/(G_0 x M_{r.} ammonia)$$

Where G is the weight of samples after adsorption (g), G_0 is the weight of samples before adsorption (g), and M_r.ammonia is the molecular weight (g/mol).

3. Results and Discussion

3.1. Crystallographic Characterization

The XRD pattern of NiO/C (figure 1) showed the success of the nickel impregnation process on carbon as supporting material. The peak at 2θ = 37.07° and 43.05° was characteristic to the presence of NiO correspond to rhombohedral structure (JCPD: 44-1159), d_{hkl} [2 0 0] and [1 1 1], respectively. The three broads peaks at these patterns similar to the precursor carbons material from coconut shells of Malaysian variety [22]. Typically, these peaks are amorphous carbon.

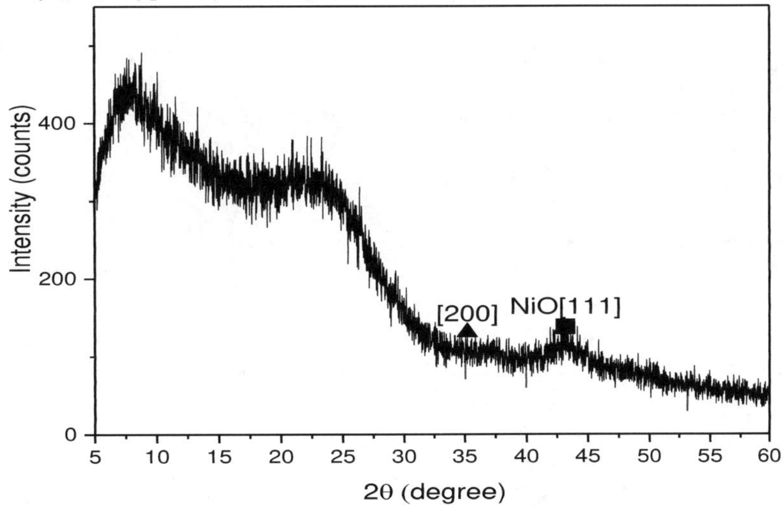

Figure 1. XRD pattern of NiO/C catalyst

3.2. Infrared Analysis

Figure 2 showed the presence of oxygen-containing functional groups of the NiO/C, after and before ammonia absorption. The existence of O-H stretching vibration was observed at the broadband of 3448 cm^{-1} wavenumber. The presence of asymmetric and symmetric C-H stretching were shown at a weak peak at 29924 cm^{-1} and 2862 cm^{-1}. The band at 1635 cm^{-1} was associated with aromatic C=C stretching present in olefinic vibrations in the aromatic region. The existence of the band in 1442 cm^{-1} assigns the presence of aromatic C, indicative of lignin C=C [23]. The band at 1118 cm^{-1} was corresponding to C-O-C stretching. Aromatic C—H bending can be found at 871 cm^{-1} [24].

IOP Conf. Series: Materials Science and Engineering 1232 (2022) 012001 doi:10.1088/1757-899X/1232/1/012001

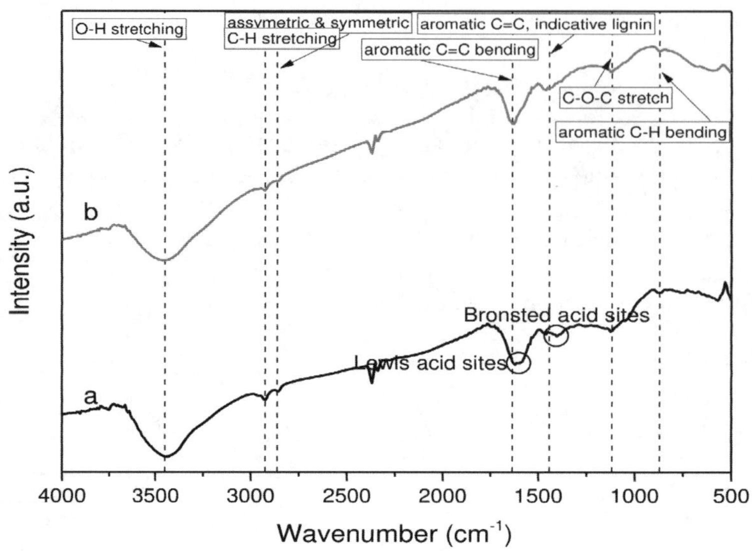

Figure 2. FTIR spectra of NiO/C: a) NiO/C after ammonia absorption and b) NiO/C before ammonia absorption

IR spectra can be used quantitively to find out the existence of Lewis and Brönsted acid sites. Figure 2 showed the changes occurring in the absorption bands, a widening at 1604 cm^{-1} region a weak peak after the absorption of ammonia gas. It is indicated ammonia coordinately bonded to Lewis acid site [25]. The band at 1404 cm^{-1} corresponded to Brönsted acid sites, which is the characteristic peak of an ammonium ion and N-H bending vibration region[26].

3.3. Acidity Characterization

The determination of the acidity of NiO/C sample was done quantitatively by the absorption of ammonia gas method. The acidity value of NiO/C samples was 3.05 mmol/g, , as presented in table 2. The presence of metals in supporting materials has been shown to contribute to the acidity value of the material [13]. The oxide form of nickel (as Lewis acid site) provides the orbitals to accept an electron from the NH_3 molecule as the basic adsorbate.

3.4. Morphology and Metals Content Analysis

The morphology features of NiO/C analysis was carried out by SEM analysis. Figure 3 showed the image of NiO/C with magnification 1000x and distribution of carbon, oxygen, and nickel contained in NiO/C sample. This image presented a large number of different sizes of pores on the surface NiO/C. The pores showed the pattern of the overlapping honeycomb-shaped tunnel, with cross-connected orientation. The capture of the content of carbon, oxygen, and nickel showed distribution uniformly on the NiO/C surface.

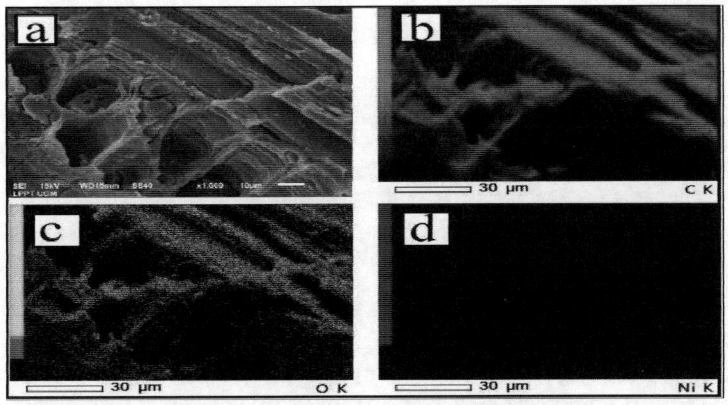

Figure 3. SEM Image of a) NiO/C, b) mapping of carbon, c) mapping of Oxygen, d) mapping of nickel

The elemental analysis of samples was determined by EDX, as shown in figure 4 and table 1. The result was confirmed by small amounts of nickel, almost of the nickel was in oxide form in the NiO/C structure.

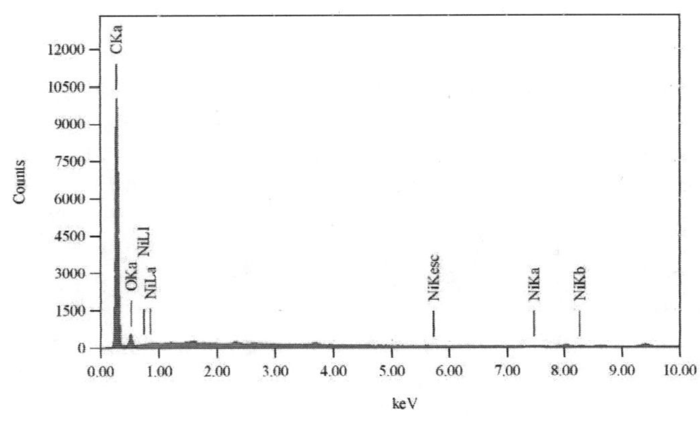

Figure 4. EDX spectra of NiO/C

Table 1. Elemental analysis based on EDX of NiO/C

Sample	Element (wt.%)			Oxide
	C	O	Ni	NiO
NiO/C	99.4	0.13	0.47	0.6

3.5. Pore Structures Analysis

The pore size distribution and average pore diameter analyzed using the N_2 adsorption data with the Barrett-Joyner-Halenda (BJH) method. Table 2 showed that the average pore diameter of NiO/C was 2.083 nm and figure 5a presented that most of the mesoporous were in the pore size range of 3-4 nm. Figure 5b showed the adsorption-desorption isotherm of the NiO/C, indicated that was of type II according to IUPAC classification, characteristic for nonporous and macroporous materials. Based on these data, the NiO/C has the characteristic pore size predominantly non-porous and macroporous-mesoporous. The synthesized NiO/C exhibited a high surface area of 520.407 (m^2/g), which has the potential in catalyst applications, as well as an adsorbent.

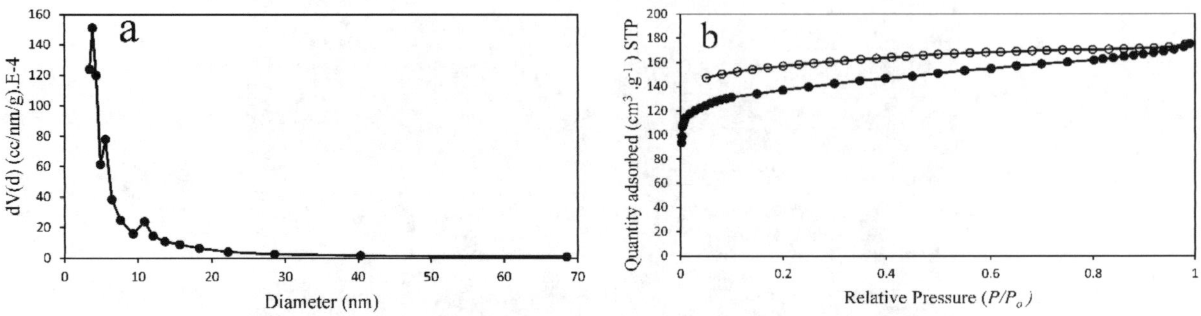

Figure 5. (a) The pore size distribution of NiO/C. (b) N_2 adsorption-desorption isotherm of NiO/C. Filled and empty symbols represent the adsorption and desorption branches.

Table 2. Surface properties analysis of NiO/C

Sample	Surface area (m^2/g)	Pore diameter (nm)	Pore volume (cm^3/g)	Acidity (mmol/g)
NiO/C	520.407	2.083	0.271	3.05

4. Conclusion

In this study, the synthesis of support material of first-row transition metal derived from saw waste of Merbau wood is provided. The synthesized material has potential acid sites and surface characteristics for the application of catalyst materials. The synthesized NiO/C has pore distribution in the meso-sized range, with several mixtures of non-porous and macropore characters. All impregnated Ni metals in the material were characteristic to the presence of oxides correspond to the rhombohedral structure.

References

[1] Qian K, Kumar A, Zhang H, Bellmer D, Huhnke R 2015 *Renew. Sustain. Energy Rev.* **42** 1055–-64
[2] Ahmed S, Ramli A, Yusup S, Farooq M 2017 *Chem. Eng. Res. Des.* **122** 33–42
[3] Maia D A S, de Oliveira J C A, Nazzarro M S, Sapag K M, López R H, de Lucena S M P, de Azevedo D C S 2018 *Chem. Eng. Res. Des.* **136** 753–60
[4] Lua A C, Lau F Y, Guo J 2006 *J. Anal. Appl. Pyrolysis.* **76** 96–102
[5] Cronje K J, Chetty K, Carsky M, Sahu J N, Meikap B C 2011 *DES.* **275** 276–84
[6] Guo J and Lua A C 2002 *Mater. Lett.* **55** 334–9
[7] Mohd W, Wan A, Shabuddin W, Ali W 2004 *Bioresour. Technol.* **93** 63–9
[8] Girgis B S, Ishak M F 1999 *Mater. Lett.* **39** 107–14
[9] Gaskin J W, Steiner C, Harris K, Das K C, Bibens B 2008 *Trans. ASABE.* **51** 2061–69
[10] Zhang S, Dong Q, Zhang L, Xiong Y 2015 *Bioresour. Technol.* **191** 17–23
[11] Ahmad F, Daud W M A W, Ahmad M A, Radzi R 2012 *Chem. Eng. Res. Des.* **90** 1480–90
[12] Zhu M, Cai W, Verpoort F, Zhou J 2019 *Chem. Eng. Res. Des.* **146** 130–40
[13] Prasiwi A D, Trisunaryanti W, Triyono T, Falah I I, Santi D, Marsuki M F 2019 *Indones. J. Chem.* **19** 575–82
[14] Byrne J F and Marsh H 1995 *Porosity in Carbons*, ed J W Patrick (New York: Halsted Press) p 2-48
[15] Daud W M A W, Ali W S W, Sulaiman M Z 2000 *Carbon N. Y.* **38** 1925–32
[16] Grams J, Niewiadomski M, Ruppert A M, Kwapiński W 2015 *J. Anal. Appl. Pyrolysis.* **113** 557–63
[17] Bartoli M, Rosi L, Giovannelli A, Frediani P, Frediani M 2016 *J. Anal. Appl. Pyrolysis.* **120** 284–96
[18] Trisunaryanti W, Suarsih E, Triyono, Falah I I 2019 *RSC Adv.* **9** 1230–37

[19] Trisunaryanti W, Falah I I, Susanto H 2016 *Int. J. ChemTech Res.* **9** 388–97
[20] Salim I, Trisunaryanti W, Triyono, Arryanto Y 2016 *Int. J. ChemTech Res.* **9** 492–500
[21] Lu M, Liu X, Li Y, Nie Y, Lu X, Deng D, Xie Q, Ji J 2016 *J. Renew. Sustain. Energy.* **8** 053103
[22] Jain A, Jayaraman S, Balasubramanian R, Srinivasan M P 2014 *J. Mater. Chem.* A. **2** 520–28
[23] González-García P 2017 *Renew. Sustain. Energy Rev.* **82** 1393–414
[24] Qian K, Kumar A, Patil K, Bellmer D, Wang D, Yuan W, Huhnke R L 2013 *Energies* **6** 3972–86
[25] Sohn J R and Lee S H 2004 *Appl. Catal.* A Gen. **266** 89–97
[26] Xie L, Gao Q, Wu C, Hu J 2005 *Microporous Mesoporous Mater.* **86** 323–28

Acknowledgments

The authors are grateful to Ministry of Education and Culture of the Republic of Indonesia-BUDI DN and Ministry of Finance-LPDP (grant number: PRJ-1475/LPDP.4/2019) for the Research Scholarship and PDUPT (2563/UN1.DITLIT/DIT-LIT/LT/2019).

Benchmarking Naïve Bayes and ID3 Algorithm for Prediction Student Scholarship

I M K Karo[1,*] , M Y Fajari[1] , N U Fadhilah[1] and W Y Wardani[1]

[1]School of Computing, Telkom University, Bandung, Indonesia

* E-mail : ichwanulkarokaro@telkomuniversity.ac.id

Abstract. Student scholarship through Indonesian smart cards (ISC) is a cash donation scholarship for all student in range 6-21 years old as a solution of destitute child or potential dropouts . Currently the large student data stored in educational database. The data was used to determine feasible receiver ISC. Manual prediction by human take long time and potentially human error. This paper aim to predict feasible receiver ISC by using educational data mining. Source of data come from a senior high school database in Riau Province, Indonesia. In this paper we compared two algorithms (Naïve Bayes and ID3) to predict receiver ISC. Eighty percent process in this paper is data pre-processing. Information of other scholarship is the most influences variable to predict student scholarship. ID3 algorithm classification has accuracy 83 percent and f1 score 71 percent. While Naive Bayes Classification has accuracy 89 percent and f1 score 72 percent. In this case Naïve Bayes algorithm is better in ID3 algorithm.

Keywords : Educational Data Mining (EDM), Indonesian smart cards (ISC), ID3, Naive Bayes

1. Introduction

Smart Indonesia Program through Indonesian smart cards (ISC) is a scholarship to all student in range 6-21 years old that destitute family or potential dropouts from Indonesia government. School through teacher proposed feasible candidate to Indonesia government and as a feedback the government give the scholarship in form cash of money. If a school has many amount of student then teacher manually classify the receiver as a candidate. This condition is not ideal, because the process takes long time and probably subjectively classification. This would cause the target beneficiaries in less appropriately.

One of solution to solve the problem is applied educational data mining. Educational data mining is applied task of data mining on education data based. There are four task of data mining, these are classification, clustering, association rule and prediction [1]. The most power full of task mining to predict based on class is by using classification techniques.

The Naïve Bayes classifier being a technique based on probabilities is applicable to a wide range of domains according to the above literature survey. It is computationally fast, has linear complexity and does not require a large amount of training data. Furthermore the algorithm justifies the membership of an instance to a particular class with the help of probabilities. The main objective of this paper is to use data mining methodologies to determine child who receive ISC. Data mining provides many tasks that could be used to study the student scholarship parameter. There are some classifiers that determine a child receive ISC, these are income of parent (father and mother), transportation, jobs, previous scholarship and room and board. In this research, there are two classification algorithms that used to predict student scholarship, these are Decision tree (ID3) and Naive Bayes. The evaluation process for the algorithm used accuracy, precision, recall and F1 measure.

2. Related Work

Many applications of data mining have been applied in real world, such as for business intelligent, decision making, education, etc. [2]. In education sector, several researcher have been implemented data mining techniques to solve problem or find pattern on educational data, such us Jalota, Chitra and Agrawal and Rashmi (2019) use kalboard 360 dataset which lies in the domain of education and gathered using learning management system (LMS) to predict performance level of students [3] on WEKA. They used J48, SVM, Random Forest, Multilayer perceptron and Naive Bayes algorithm. The result show that Multilayer perceptron had best performance for the predict data set.

Previous research to predict student performance has been proposed Devasia, Tismy and Vinushree, TP and Hegde and Vinayak in 2016 they used Naive Bayes algorithm to predict potential dropout of student in higher education [4]. In line, Ahmed, Abeer Badr El Din and Elarab Ibrahim Sayed in 2014 also had publish their research. They used ID3 algorithm on data set from a student's database used in one of the educational institutions, on the sampling method of Information system department from session 2005 to 2010 [5].

Classification algorithm also to analyze Students' Behavior. Abu Tair, Mohammed M and El-Halees and Alaa M analyzed educational data to figure out the reasons behind students' behaviors and make a decision for solutions and treatment paths [6]. They map student behavior into 12 behaviors.

In addition, many researcher have been implement Naive Bayes algorithm to predict some real case, Naive Bayes for predicting purchase [7], Naive Bayes to predict slow learners in education sector [8].

The Naïve Bayes algorithm has been applied to educational data mining in different contexts. Therefore Naïve Bayes classifier is suitable for effective classification of student feedback for faculty on class room delivery as class room consists of smaller data set [9].

3. Design Process

Generally, data mining process start from availability of data, feature selection, pre-processing data, data mining, evaluation and the end is we have the knowledge from the data shown in figure 1.

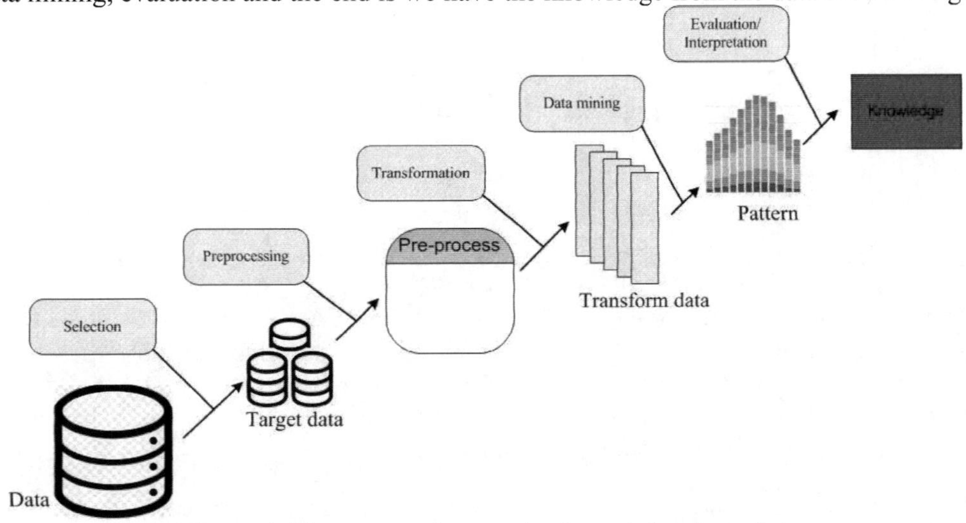

Figure 1. The stages of extracting knowledge from data

IOP Conf. Series: Materials Science and Engineering 1232 (2022) 012002 doi:10.1088/1757-899X/1232/1/012002

Flow of the research shown in figure 2 shows that 80 percent process is collecting data and preprocessing data. After we have well- form data, thus we have data train dan data test, next step is tried to modelling data by using ID3 and Naive Bayes algorithm by using data train. Next step is to evaluate the model using accuracy, precision, recall, and f1 score for both algorithms.

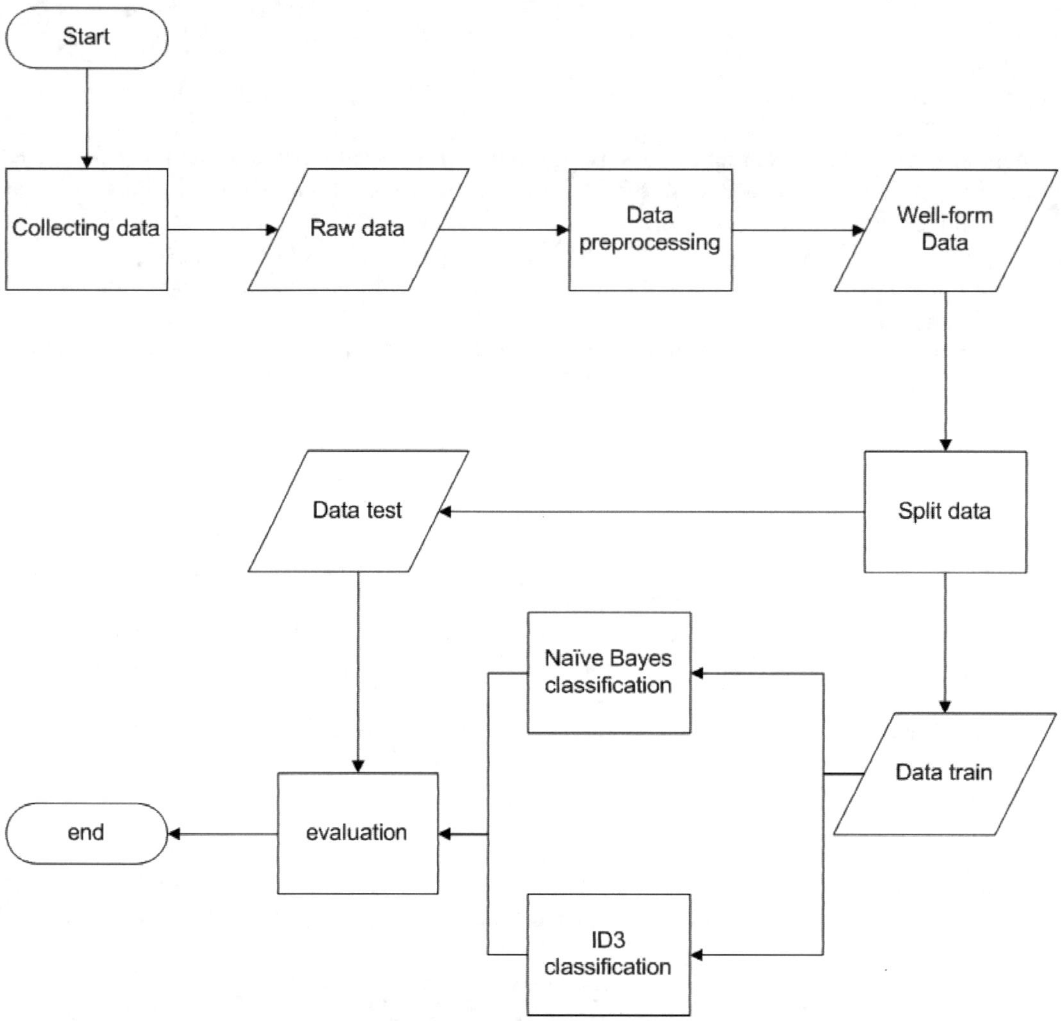

Figure 2. Flowchart of Research

3.1. Data Pre-processing

In this research, source of data come from receiver ISC data at a senior high school in Riau Province, Indonesia. Initially size of the data is 882 records, seven attributes, and two class (table 1). There are 778 class not and 104 class yes.

Data preprocessing are critical step in data mining process. In this step only those fields were selected which were required for data mining. A few derived variables were selected by using feature selection, transform the attribute into categorical type and handling the missing value. Next step is splitting data used K-fold, where k=5 into train and testing.

Table 1. Student related variable

Variables	Description	Possible value
Type of domicile	student domicile	in home, boarding school, rent boarding, stay with other family, other
Type of Transportation	student transportation go to school	public transportation, walking, private vehicle, school bus, motorcycle taxi, motorcycle, other
other scholarship	a student is receiver other/previous scholarship	yes or not
job of father	job information of father	labour, employee, government employee, farmer, entrepreneur, retirees, other
income of father	range income of father	≤1,000,000, 1,000,000 – 2,000,000, 2,000,000 – 5,000,000, ≥ 5,000,000
Job of mother	job information of mother	labour, employee, government employee, farmer, entrepreneur, retirees, other
Income of mother	range income of mother	≤1,000,000, 1,000,000 – 2,000,000, 2,000,000 – 5,000,000, ≥ 5,000,000
class	receiver ISC or not	Yes or not

3.2. Classification Method

There are many classification algorithm. Each algorithm has advantages and disadvantage. Each algorithm is not always power full to all kind of data, such us many researcher use decision tree for categorical data. A decision tree is a flow- chart-like tree structure, where the tree structure have root and leaf/child. Root/child are denoted by rectangles, it mean the attributes. While leaf are denoted by circles, it mean the decision or class. All internal nodes have two or more child nodes. All internal nodes contain splits, which test the value of an expression of the attributes [5].

3.2.1. Iterative Dichotomiser 3 (ID3): is one of existing decision tree algorithm that using Shannon Information Theory. The basic idea of ID3 algorithm is to construct the decision tree by employing a top-down, greedy search through the given sets to test each attribute at every tree node. In order to select the attribute that is most useful for classifying a given sets, we introduce a metric - information gain. To find an optimal way to classify a learning set we need some function which provides the most balanced splitting. The information gain metric is such a function. Given a data table that contains attributes and class of the attributes, we can measure homogeneity of the table based on the classes using entropy shown in equation 1

$$\boldsymbol{H\,(S) = \sum_{j=1}^{k} -P_j \log_2 P_j} \tag{1}$$

Splitting criteria used for splitting of nodes of the tree is information gain. To determine the best attribute for a particular node in the tree we use the measure called Information Gain. The information gain, Gain (S, A) of an attribute A, relative to a collection of examples S, is defined as in equation 2

$$Gain\,(S, A) = H\,(S) - \sum_{t\,\in T} p(t)\;H(t) \tag{2}$$

3.2.2. Naive Bayes: The Naive Bayes algorithm is a simple probabilistic classifier that computes a set of probabilities by summing the frequency and value combinations of the given data set [7]. Naive Bayes is a kind of classifier which uses the Bayes theorem. It predicts membership probabilities for each class such as the probability that given record or data point belongs to a particular class (equation 3). The class with the highest probability is considered as the most likely class. Naive Bayes Classifier assumes that all the features are unrelated to each other to study their individual effect on feedback. Presence or absence of a feature does not influence the presence or absence of any other feature. Though these may depend on each other and on existence of other features all these features are considered to independently contribute to the probability that it is a valid feedback. The hypothesis is tested for given multiple evidences (features).

$$\mathbf{P\,(H\,|\,X)\;=\;P\,(X\,|\,H)\,.\,P\,(H)\;P\,(X)} \tag{3}$$

3.2.3. Evaluation Method: Most important question that arises is how good is model to solve our problem. Thus to evaluated the performance of ID3 and Naive Bayes algorithm used Accuracy, Precision, Recall and F1 measure [3].

4. Result and Discussion

In pre-processing step, we remove typo of data and noise symbol, analyse variable in education database use correlation value (Table 2). Only domicile variable has linear influence with class and other variable are not. Other scholarship information is highest influence than other variable to determine receiver ISC.

Table 2. Correlation value of variable

Variable	Correlation value
Other scholarship	-0.43
Job of father	-0.071
Income of mother	-0.065
Job of mother	-0.052
Domicile	0.035
Transportation	-0.034
Income of father	-0.031

4.1. Classification using ID3 Algorithm

In this part, Data were classified by ID3 algorithm and running in Python. Generally, ID3 algorithm able to predict feasible student scholarship with accuracy 83 percent. In more, performance of ID3 algorithm shown on Table 3. the algorithm has good result for predict not class, shown on high precision, recall and F1 measure.

However the algorithm has poor result for predict yes class, because value of precision, recall dan F1 measure are under 50 percent

Table 3. Performance of ID3 algorithm

	precision	recall	F1
Not	93	93	93
Yes	50	43	49

4.2. Classification using Naïve Bayes

Next experiment, Data were classified by Naive Bayes algorithm and running in Python. Naïve bayes algorithm able to predict feasible ICS receiver with accuracy 89 percent. Performance of the algorithm shown on table 4. The algorithm performance is not significant different with ID3 algorithm. Classification using Naive Bayes has high precision, recall and F1 to predict the class. High precision show that the algorithm correctly predict the positive class for ISC receiver data. In addition model classification able to observe actual class. However, researcher analyse that imbalance data has not been well handle. Thus precision, recall and F1 for class not is highest than class yes.

Table 4. Performance of Naive Bayes algorithm

	precision	recall	F1
Not	92	95	94
Yes	56	43	50

4.3. Compared both algorithm

Based on both experiment, both algorithm able to predict feasible ISC receiver. Although performance of both algorithm is similar. But based on four parameters performance, Naïve bayes algorithm is higher than ID3 (table 5). In other word Naive Bayes algorithm is better than ID3 for classifying receiver ISC.

Table 5. Compared both algorithm

	precision	recall	F1	Accuracy
ID3	71.5	68	71	83
Naïve Bayes	74	69	72	89

5. Conclusion

In this paper, There are seven variable in classification process and the most influences is other scholarship to predict feasible ISC receiver. Naïve Bayes and ID3 algorithm able to predict receiver ISC, which data come from educational databased. Naïve Bayes is batter than ID3 to predict receiver ISC. This study will help school to determine receiver ISC in quickly. Our next hypothesis is handling imbalance data will increase the performance both algorithms

References

[1] Jiawei Han, Jian Pei, and Micheline Kamber. *Data mining: concepts and techniques*. Elsevier, 2011.

[2] Dursun Delen. *Real-world data mining: applied business analytics and decision making*. FT Press, 2014.

[3] Chitra Jalota and Rashmi Agrawal. Analysis of educational data mining using classification. In *2019 International Conference on Machine Learning, Big Data, Cloud and Parallel Computing (COMITCon)*, pages 243–247. IEEE, 2019.

[4] Tismy Devasia, TP Vinushree, and Vinayak Hegde. Prediction of students performance using educational data mining. In *2016 International Conference on Data Mining and Advanced Computing (SAPIENCE)*, pages 91–95. IEEE, 2016.

[5] Abeer Badr El Din Ahmed and Ibrahim Sayed Elaraby. Data mining: A prediction for student's performance using classification method. *World Journal of Computer Application and Technology*, 2(2):43–47, 2014.

[6] Mohammed M Abu Tair and Alaa M El-Halees. Mining educational data to improve students' performance: a case study. *Mining educational data to improve students' performance: a case study*, 2(2), 2012.

[7] Fitriana Harahap, Ahir Yugo Nugroho Harahap, Evri Ekadiansyah, Rita Novita Sari, Robiatul Adawiyah, and Charles Bronson Harahap. Implementation of na¨ıve bayes classification method for predicting pur- chase. In *2018 6th International Conference on Cyber and IT Service Management (CITSM)*, pages 1–5. IEEE, 2018.

[8] Parneet Kaur, Manpreet Singh, and Gurpreet Singh Josan. Classification and prediction based data mining algorithms to predict slow learners in education sector. *Procedia Computer Science*, 57:500–508, 2015.

[9] Sandhya Maitra, Sushila Madan, Rekha Kandwal, and Prerna Mahajan. Mining authentic student feedback for faculty using na¨ıve bayes classifier. *Procedia computer science*, 132:1171–1183, 2018

The Acceptance of Primary Health Centre Information System Among Health Staff: An extended TAM Model

Laili Rahmatul Ilmi

Universitas Jenderal Achmad Yani
Jl. Brawijaya, Ring Road Barat, Ambarketawang, Gamping, Sleman, Yogyakarta, Indonesia

E-mail: lailiilmi@gmail.com

Abstract. Ngaglik I Public Health Center in Sleman is one of the health centers that have already applied SIMPUS (an application of management information system for public health center). The implementation of clinical information systems could potentially improve the performance of health care facilities, save operational costs, and increase patient satisfaction. In fact, the clinical information systems applied by public health center vary. One of them is SIMPUS. However until present, there have not been any assessments of the acceptance of the clinical information system, especially SIMPUS, at Ngaglik I Public Health Center. To describe the use of SIMPUS at Ngaglik I Public Health Center and evaluate user understanding in accepting SIMPUS application based on the Technology Acceptance Model (TAM). This was a quantitative study which used a survey method and a cross-sectional design. The research sample consisted of 50 respondents who were selected using a simple random sampling method. The data analysis technique was the SEM-PLS analysis with the TAM model and SmartPLS software. Statistically, the relationship between the Job Relevance and Perceived Ease of Use variables was weak (p-value = 0.818). There was a weak relationship between Output Quality and Perceived Ease of Use (p-value = 0.208). There was a strong relationship between Result Demonstrability and Perceived Ease of Use (0.000). There was a strong relationship between Result Demonstrability and Intention to Use (p-value = 0.011). There was a weak relationship between Screen Design and Intention to Use (p-value = 0.870). There was a weak relationship between Perceived Ease of Use and intention to use. Users' job relevance to SIMPUS did not have any effect on perceived ease of use. Output quality did not have any effect on perceived ease of use. The screen design of SIMPUS display had no effect on intention to use. Intention to use did not affect system use. Intention to use did not affect system use. Facilitating condition did not have any effect on system use. System use was influenced by perceived ease of use and intention to use. Perceived ease of use and intention to use were influenced by result demonstrability.

Keywords: Acceptance, information system, public health center, TAM

1. Introduction

This study measured SIMPUS acceptance model at Ngaglik I Public Health Center. The researcher evaluated the acceptance using several variables, including Job Relevance, Output

Quality, Result Demonstrability, Screen Design, Perceived Ease of Use, Intention To Use, Facilitating Conditions, and System Use. SIMPUS acceptance was measured using the TAM model. This theory was originally developed by Davis [1] This conceptual concept will continue to be developed in line with the needs and findings of previous studies. The utilization of information systems in the clinical sector is very useful to improve service performances [2], accelerate performance, and support decision making [3] Primary health centers certainly require regional health records because this health center provides basic health services, one of which is related to coverage or needs for monitoring data [4]. Information technology really helps all aspects of activities. One of which is in the health sector, i.e. to create integration with the social security system, one of which is BPJS (Healthcare and Social Security Agency). Ngaglik I Health Center has applied SIMPUS since 2014, but until now, not all units at this health center have had good acceptance of SIMPUS. Several previous studies showed that the acceptance of an integrated clinical information system accelerates services and significantly supports policymakers in making decisions [5][6]. The TAM theory significantly explains that the key to the success of system implementation is human factors [7], but other studies explain that the key to success is due to the display of the information system and integrated system [8].

2. Application of Clinical Information System

Assessment of the benefits of information system implementation was first developed by McLean and Delon. This model, which comprises of six components, is the key concept that is important for measuring, analyzing, and reporting the success of IS since the development of the system, containing various features and quality of IS. Further, the theoretical framework was adopted [1] using the Macro, Meso, and Micro level analysis. The following figure is the evaluation model developed by Delon and McLean. In fact, several previous studies have focused on the application and evaluations of information systems. Several other studies have also developed similar terms [9], while other studies focused on the evaluation of end users [2] and studies which measured the effect of the application on electronic medical records [3], other researchers explained that the results of EMR data to be reviewed are 1) quality, i.e. the data contained in the records could provide useful and meaningful information; 2) complete, i.e. all items that must be made available are meaningful and able to provide information for other users; 3) by following what has been done, so the data could provide useful information when a user reads the data which have been recorded.

In the above theoretical framework, McLean and Delon revealed that MIS successfully brings positive effects on organizations, thus bringing effects on individuals as well. In order to ensure that management information systems have an effect on individuals, user satisfaction is also measured, so the achievement of both covers the achievement of system quality and information quantity [4]. The above theoretical framework shows that there are six key components, including system quality components and information quality of the system. Both system quality and information quality influence users and user satisfaction, so hopefully IS will bring a positive effect not only on individuals but also on organizations.

Further, the above theoretical framework was developed into three dimensions including the macro, meso, and micro dimensions [5]. The macro dimension consists of four main variables, namely a) the objectives of evaluating health standards used in the information system, b) the financing or incentives earned, c) bridging profit regulation, d) policies, the legal aspect used to implement the system. The meso dimension is used to evaluate the extent to which the capacity is adopted. This dimension consists of three main components, namely a) individuals or groups; b) organization, i.e. how it is implemented in accordance with the organization's strategy, culture, and socio-economy, and c) the implementation has an effect on every level. In addition, the micro dimension is used to evaluate the quality of both information and system being used. The micro dimension consists of three main components, namely: a) the quality of the clinical information system to measure the accuracy of clinical data, the completeness of clinical data, and the availability of clinical information, display, system security and responses which support system, b) the quality of users' intensity in using the information system, ease of system, user satisfaction with the ease of information system, and user competence, c) benefits or net benefits obtained from service quality; access, i.e. ease of access to services provided by service providers and productivity of users or service providers to exchange patient medical records with coworkers to improve patient safety.

Technology Acceptance Model (TAM) theory was developed by Davis (1989). This theory comprises of two main constructions, namely: a) Perceived usefulness is the level of belief that using technology will improve performance. If someone believes that an information system is useful, s/he will use it, and the opposite. Perceived usefulness can be defined as making work easier, useful, productivity, effectiveness, and work performance [3]; and b). Perceived ease of use is one's level of confidence that using technology is easy. Davis (1989) provided some indicators for perceived ease of use, including systems that are easy to learn, easy to use and easy to control, and the systems make it easy for users to become skillful in using the system.

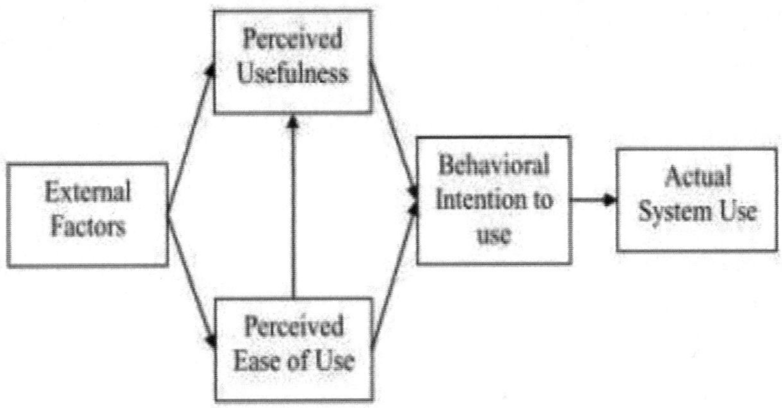

Figure 1. Theory *Technology Acceptance Model* (TAM)

3. Method

This was a quantitative research using a survey method and analytical correlational research design. The sample consisted of 50 respondents who were selected by using a simple random sampling method. The data analysis technique was the SEM-PLS analysis technique with the TAM model and SmartPLS software. The objective of this study was to measure SIMPUS acceptance by the health workers at Ngaglik I Health Center. From the 50 respondents, only 32 of them could be proceeded because the remaining 18 were neither returned nor completed. The data processing was carried out in four stages. First, the validity and reliability of the questionnaire results were tested, in addition to measuring the characteristics of the respondents to be presented in tabulations (Table 1). Second, the measured variables were then entered into the SmartPLS software to analyze the TAM model. The data analysis was carried out in two stages: analysis using model testing to calculate its validity and reliability and analysis using structural measurement to test the hypothesis/relationship between variables. Third, testing the measurement between variables. Fourth, structural measurement between variables.

4. Results

4.1 Characteristics of Respondents

Table 1 shows the characteristics of respondents.

Table 1. Characteristics of Respondents

Item	Frec	(%)
Sex		
Male	5	15.63
Female	27	84.38
Education		
Senior high school	5	15.63
Vocational school	5	15.63
Undergraduate	22	68.75
Job description		
Administration staffs	5	15.63
Doctors	3	9.38
Pharmacist	1	3.13
Laboratory technician	1	3.13
Medical record technicians	2	6.25
Midwives	8	25.00
Nurses	6	18.75
Nutritionist	1	3.13
Admission staffs	4	13.51

4.2 Analysis of outer model

The questionnaires were analyzed using the SmartPLS 3.2.3. The data analysis was carried out in two stages: first, analysis using model testing to calculate its validity and reliability, second, analysis using structural measurement to test the hypothesis/ relationship between variables. The questionnaires adopted from Delon and McLean's theoretical framework were distributed to the health worker respondents who used electronic medical records to measure their perceptions (the questionnaires are shown in appendix). The questionnaires were filled out by 32 health worker respondents who were willing, including doctors, nurses, medical record technicians, laboratory technician, radiology technicians, and pharmacists. The results of the validity and reliability assay are shown in Table 2.

Table 2 showed that the outer model values were all valid based on the loading factors. The reliability was also good. Based on these results, it can be concluded that the questionnaires were valid and reliable to make a model. The following stage was to model the perceived effect of information system implementation and information quality on customer satisfaction at Ngaglik I Health Center. The modeling aimed to support qualitative data to determine user understanding in accepting the SIMPUS application based on the Technology Acceptance Model (TAM).

IOP Conf. Series: Materials Science and Engineering 1232 (2022) 012003 doi:10.1088/1757-899X/1232/1/012003

Table 2. Item Loading, Composite Reliability, Cronbach Alpha and AVE

Factor	Item	Loading	Composite reliability	Cronbach Alpha	AVE
FC	FC1 FC2 FC3	0.963 0.944 0.620	0.889	0.836	0.734
ITU	ITU1 ITU2 ITU3	0.927 0.934 0.990	0.966	0.947	0.904
JR	JR1 JR2	0.803 0.952	0.873	0.736	0.776
OQ	OQ1 OQ2	0.952	0.945	0.828	0.895
PEOE	PEOE 1 PEOE 2	0.894 0.885	0.884	0.737	0.792
RD	RD1 RD2	0.914 0.933	0.920	0.828	0.853
SD	SD1 SD2	0.945 0.849	0.893	0.773	0.807
SU	SU1 SU2	0.877 0.916	0.981 0.984	0.964	0.965

JR = Job Relevance *PEOE = Perceived Ease of Use*

OQ = Output Quality *ITU = Intention to Use*

RD = Result Demonstrability *SU = System Use*

SD = Screen Design *FC= Facilitating Condition*

4.3 Analysis results of the effects between variables

To test the effect of variables was based on the results of calculations through the bootstrapping method on the SmartPLS 3.2.3. The data were analyzed with a significance level (α) = 5% using a two-tailed test with degrees of freedom (df) = (n-k) = (100-5) = 95. The analysis results are as follows.

This study had eight hypotheses (Table 3), namely:

1) There is a significant relationship between Job Relevance (JR) and Perceived Ease of Use (PEOU)
2) There is a significant relationship between Output Quality (OQ) and Perceived Ease of Use (PEOU).
3) There is a significant relationship between Result Demonstrability (RD) and Perceived Ease of Use (PEOU).
4) There is a significant relationship between Result Demonstrability (RD) and Intention to Use (ITU).
5) There is a significant relationship between Screen Design (SD) and Intention to Use (ITU).
6) There is a significant relationship between Perceived Ease of Use (PEOU) and System Use (SU).
7) There is a significant relationship between Intention to Use (ITU) and System Use (SU).
8) There is a significant relationship between Facilitating Condition (FC) and System Use (SU).

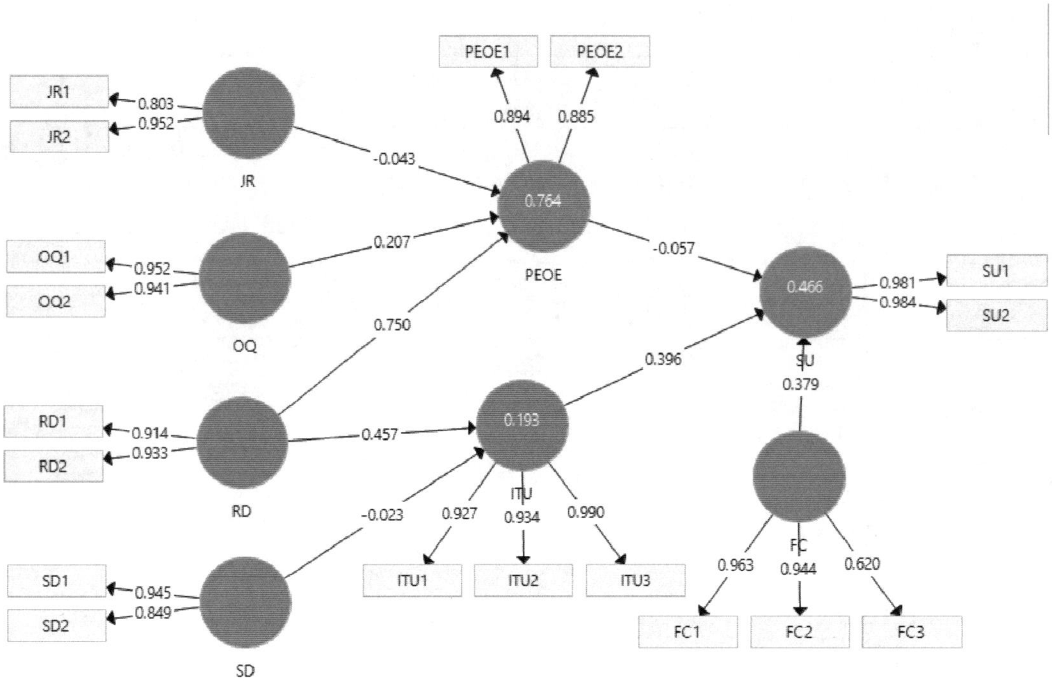

Figure 2. Results of Data Influence between Variables

JR = Job Relevance
OQ = Output Quality
RD = Result Demonstrability
SD = Screen Design
PEOE = Perceived Ease of Use
ITU = Intention to Use
SU = System Use
FC= Facilitating Condition

Table 3. Test results for the structure model between variables

Hypothesis	*Path*	*Standard Deviation (STDEV)*	*Path Coefficient*	*T Statistic (|O/STDEV|)*	*P Value*	*R square*
H1	*JR→ PEOE*	0.189	-0.038	0.230	0.818	Rejected
H2	*OQ → PEOE*	0.164	0.220	1.260	0.208	Rejected
H3	*RD→PEOE*	0.214	0.725	3.504	0.000	Accepted
H4	*RD→ITU*	0.399	0.399	0.307	0.011	Accepted
H5	*SD→ITU*	0.144	0.015	0.163	0.870	Rejected
H6	*PEOE→ SU*	0.124	-0.073	0.461	0.645	Rejected
H7	*ITU→ SU*	0.212	0.391	0.641	0.063	Rejected
H8	*FC→ SU*	0.229	0.411	1.656	0.098	Rejected

The correlation coefficient for a valid hypothesis was the correlation between Result Demonstrability (RD) and Perceived Ease of Use (PEOU), i.e. 0.725, meaning that there was a very strong relationship between the two variables. The correlation coefficient between Facilitating Conditions (FC) and System Use (SU) was 0.411, meaning that there was a quite strong relationship between the two variables. On the other hand, the correlation coefficient between Job Relevance (JR) and Perceived Ease of Use (PEOU) was -0.038, meaning that there was a weak relationship between the two variables.

5. Discussion

Validity assay is carried out to test the accuracy of a measuring instrument towards any objectives to be achieved. A measurement result is said to be valid if it matches the objectives, and vice versa. The indicators for the level of validity generally use a rule of thumb to assess convergent validity and discriminant validity of measurement instruments. A convergent validity parameter is loading factor greater than 0.7 for confirmatory studies and loading factor ranging from 0.6 to 0.7 for exploratory studies. Another parameter is that the Average Variance Extracted (AVE) has to be greater than 0.7. Nevertheless, a loading factor of 0.5-0.6 is still sufficient for the early stage of measurement scale development research [4]. The results of the validity assay using the SmartPLS 3.2.3 software indicated that the measurement model met the aspects of both convergent validity and discriminant validity. This means that the measurement model was valid, evident from the correlation coefficient between variables and constructs (question items in the questionnaire) and the Average Variance Extracted (AVE) for each variable greater than 0.5.

In the derivative of the Job Relevance variable as shown in the figure 2, it can be seen that the correlation coefficient between Job Relevance and JR1 is 0.803 and JR2 is 0.952. This means that the respondents perceived that SIMPUS was used in accordance with the jobs, where staffs and technicians could easily fill in administrative data and/or clinical data into the system. This way, ease of use can improve service performance and employee performance. This is in line with a research conducted by [1] that using clinical information systems in primary health center helps speed up the registration process and makes it easier for administrative staffs to fill in data as well as exchange information within the units [6].

As seen in the figure 2, the Output Quality variable comprises of OQ1 and OQ2, and it can be seen that the correlation coefficient between OQ and OQ1 is 0.952 and OQ2 is 0.941. The respondents perceived that using SIMPUS could help them satisfactorily complete their works with a good quality. This is in line with a research conducted by [7] showing that the use of clinical information systems speeds up works, improves service quality, saves time, and speeds up services to users compared to paper-based services. In addition, the use of clinical information technology improves the speed and quality of health service provision, as well as lowers operational costs [8].

The derivatives of the Result Demonstrability variable are RD1 and RD2 and as seen in the figure 2, it can be seen that the correlation coefficient between RD with RD1 is 0.914 and with RD2 is 0.933. The respondents perceived that the results obtained by using SIMPUS were highly accurate and SIMPUS made it easy for them to exchange information with other users. This is in line with another research showing that adopting a clinical information system is very useful because it facilitates the exchange of information [9]. The Screen Design variable has two constructs namely SD1 and SD2 5. In the Perceived Ease of Use variable, there are PEOE1 and PEOE2. Based on the figure 2, it can be seen that the correlation coefficient between PEOE with PEOE1 is 0.894 and with PEOE2 is 0.885. The respondents perceived that using SIMPUS made it easy for them to obtain data due to integrated database within their service units. This is in line with a research conducted by [8] explaining that the data inputted into the clinical information system greatly enhances accuracy, facilitates the process of reading data related to medical history including treatment history. 6. In the Intention to Use variable, there are ITU1, ITU2, and ITU3. Based on the figure 2, it can be seen that the correlation coefficient between ITU with ITU1 is 0.927, with ITU2 is 0.934 and with ITU3 is 0.990. The respondents perceived that since SIMPUS application started to be used, the health workers were very enthusiastic about using it and planned to use it regularly. The results of some previous studies showed that human resources, who regularly use clinical information systems to work on their jobs, experience an increased work productivity [9]. Besides, the use of the system could accelerate work completion, facilitate each job, help minimize cost and save time [10]. 7. In the Facilitating Condition variable, there are FC1, FC2, and FC3. Based on the figure 2, it can be seen that the correlation coefficient between FC with FC1 is 0.963, with FC2 is 0.944 and with FC3 is 0.001. The respondents perceived that using SIMPUS has made it easier for them to complete patient data and the systems were easy to use. Hopefully SIMPUS will improve not only the performance of users in the registration unit, but also ease in information exchange, accuracy in data completion to allow for the data to be used as comprehensive information [9].

6. Conclusion

From the system using derivatives SU1 and SU2, it can be seen that the correlation coefficient between Su and SU1 is 0.981 and SU2 is 0.984. The respondents perceived that they consciously used SIMPUS to work on their daily works, thus improving user performance. A research by [10] showed that the effects of implementing clinical information systems for health care providers are: speed up work, allow the process of inputting data to be more effective and efficient, and help data processing and support decision making.

Acknowledgments
Thank you very much for all health practioner Ngaglik I primary health center, Medical Record and Health Information System Departmen, Faculty of Health University of Jenderal Achmad Yani Yogaykarta and our tim research Risa Milleniasari and Noor Tata Afni.

7. References

[1] Vest JR, Kern LM, Silver MD, Kaushal R. The potential for community-based health information exchange systems to reduce hospital readmissions. 2014;1–6.
[2] Taylor P, Holden RJ. Behaviour & Information Technology A theoretical model of health information technology usage behaviour with implications for patient safety. 2009;(April 2013):37–41.
[3] Saleem JJ, Plew WR, Speir RC, Herout J, Wilck NR, Marie D, et al. Understanding barriers and facilitators to the use of Clinical Information Systems for intensive care units and Anesthesia Record Keeping : A rapid ethnography. Int J Med Inform [Internet]. 2015;84(7):500–11. Available from: http://dx.doi.org/10.1016/j.ijmedinf.2015.03.006
[4] Angga Eko Pramono. Penerimaan Sistem Informasi Manajemen Puskesmas (SIMPUS) Oleh Tenaga Kesehatan Di Wilayah Kota Yogyakarta. Yogyakarta; 2016.
[5] Lau F, Price M, Boyd J, Partridge C, Bell H, Raworth R. Impact of electronic medical record on physician practice in office settings : a systematic review. 2012;
[6] Nistor GC. An extended technology acceptance model for marketing strategies in social media. Rev Econ Bus Stud. 2019;12(1):127–36.
[7] Zaied ANH. An Integrated Success Model for Evaluating Information System in Public Sectors. J Emerg Trends Comput Inf Sci. 2012;3(6):814–25.
[8] Ahani A, Nilashi M, Ahmadi H. Journal of Soft Computing and Decision Support Systems Evaluating the Barriers of Hospital Information System Implementation Using Analytic Network Processes (ANP) Method. 2016;3(4):30–8.
[9] Hillestad R, Scoville R, Taylor R. Can Electronic Medical Record Systems Transform Health Care ? Potential Health Benefits , Savings , And Costs. 2016;(November).
[10] Dube MB, Mlotshwa PR, Africa S, Dube M. Factors influencing enrolled nursing students ' academic performance at a selected private nursing education institution in KwaZulu-Natal. 2017;1–7.

Prospect for Designing Wearable Devices for Children's Asthma Attack Detection Based on Internet of Things and Cloud Computing

A. I. Wicaksono [1], R Sahtyawan [1], D Hariyadi [1] and F E Nastiti [2]

[1]Universitas Jenderal Achmad Yani Yogyakarta, Indonesia
[2] Universitas Duta Bangsa Surakarta, Indonesia

Email: [1] ariefikhwanwicaksono@gmail.com, [2] ramasahtyawan@gmail.com, [3] dedy@unjaya.ac.id, [4] faulinda.en@gmail.com

Abstract. Indonesia's geographical location as an archipelago makes this country has a humid tropical climate. This climate will affect asthmatics. Asthma attacks a patient suddenly, this is very dangerous especially if the patient is a child. Before attacking usually, asthma will react to the body with the characteristics of changes in breathing frequency, pulse frequency, mild, severe, even to the point of stopping breathing. Asthma attacks will be very dangerous for children who are not under the supervision of their parents (being in school for example). It can be imagined if asthma with severe clusters aims. The purpose of this study is to propose the design of wearable-type devices used in children at home to facilitate monitoring of asthma attacks through the internet network.

1. Introduction

The population of Indonesia, which reaches more than 260,000,000 people, has a diverse ethnic, religious, ethnic, cultural and geographical location that is characteristic [1]. That much population does not have national data related to asthma in childhood with review of emergency visits, hospitalization, and death in a hospital. Then the International Study of Asthma and Allergies in Childhood (ISAAC) conducted a study on determining the prevalence of asthma in children in several provincial capitals, see **Table 1** [2].

Disruption of the respiratory system, asthma sufferers will feel shortness of breath and coughing this is due to lack of air. Asthma attacks can occur at night or early morning [3 Globally, Indonesia is ranked 20th for asthma-related deaths. About one in 22 people suffer from asthma. However, only 54% were diagnosed with 30% of cases well controlled [4]. The results of the Basic Health Research (Riskesdas) in 2018 showed an average result of the prevalence of asthma in Indonesia of 2.4%. The number of patients with asthma in outpatients aged <4th year as much as 15.42% [5].

The National Child Asthma Guidelines issued by the Indonesian Pediatrician Association in 2015 classified asthma attacks into three: Moderate mild asthma attacks, severe asthma attacks, and Asthma attacks with the threat of stopping breathing, see **Table 2** [6]. Based on **Table 2** Asthma attack symptoms can be detected using sensors known as body sensor networks (BSN). In BSN there are several sensors such as heart rate sensor, finger sensor, oximeter sensor, and peak expiratory flow meter [7][8].

Table 1. Epidemiology Studies on Childhood Asthma in Indonesia (2008 - 2017)

Capital of the Province	Year	Age (years)	Number	Prevalence
Bandung	2012	7 - 14	332	9.6%
Jakarta	2008	13 - 14	10273	12%
	2010	13 - 14	2023	13%
	2011	13 - 14	562	9.4%
	2011	13 - 14	2003	6.4%
Medan	2012	9 - 12	238	9.7%
Padang	2009	6 - 7	933	8.0%
Palembang	2008	6 - 7	1026	8.0%
	2017	6 -7	4007	4.2%
Semarang	2013	13 -1 4	900	7.1%
Yogyakarta	2016	6 - 7	2106	11%
		13-14	3445	12%

Table 2. Asthma Attack Classification

Light	Heavy	Stop Breathing Threat
• Can speak in sentence form	• Speak in a few words	• Sleepy
• Prefer to sit rather than lie down	• Sit on your arms	• Letargi
• There are no signs of anxiety	• There are no signs of anxiety	• No breath sounds
• Increased breathing frequency	• Increased breathing frequency	
• Pulse rate increases	• Pulse frequency increases	
• Oxygen saturation level (SpO2) 90%-95%	• Oxygen saturation level (SpO2) <90%	
• Peak Expiratory Flow (PEF) level > 50%	• Peak Expiratory Flow Levels (PEF) $\leq$ 50%	

The value of Peak Expiratory Flow can indicate a person's health. Using the MPX 5100 sensor mounted on a pipe orifice plate can measure the air pressure of a person's breath so that asthma attacks can be detected [9]. However, this research does not include wearable-type devices. Photoplethymography (PPG) sensor that can measure a person's pulse and oxygen saturation level. Determination of the location of PPG sensors greatly affects the accuracy of the placement of sensors with wearable-type must be precise [10]. There are also many similar studies that measure the pulse using a heart rate sensor and oxygen saturation level using an oximeter whose data is sent to the monitor system via a 2.4 GHz frequency. This system is ideally installed in a hospital or clinic because it cannot be categorized as a wearable-type device. Therefore in this paper we propose the design of wearable-type devices used in children at home to facilitate monitoring of asthma attacks through the internet network.

2. Related Works

Internet of Things (IoT) is a new paradigm in communication networks, the data generated is processed in various ways including using certain algorithms [11] [12]. Cisco Systems proposes a model called the Internet of Things Reference Model with seven layers: Physical Devices & Controllers, Connectivity, Edge (Fog) Computing, Data Accumulation, Data Abstraction, Application,

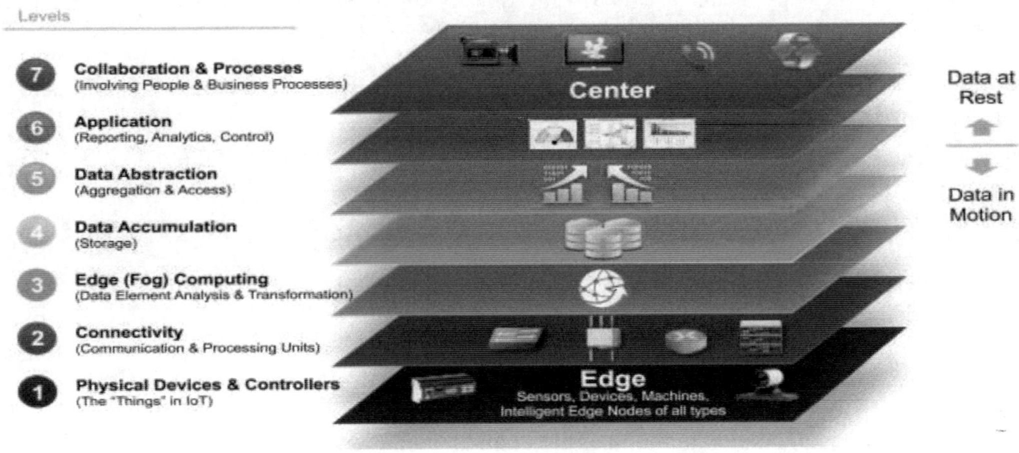

Figure 1. Internet of Things Reference Models

and Collaboration & Processes, see Figure 1 [12].

Other research has also discussed the IoT e-Health Framework for monitoring asthma attacks consisting of three layers: patient, fog computing, and cloud computing, see Figure Figure 4. At the patient layer and around attached several sensors related to health, fog computing processes data from sensors in patients to help diagnose signs of disease, and cloud computing is a layer for storing data and collaborative analysis with various parties including the family [13].

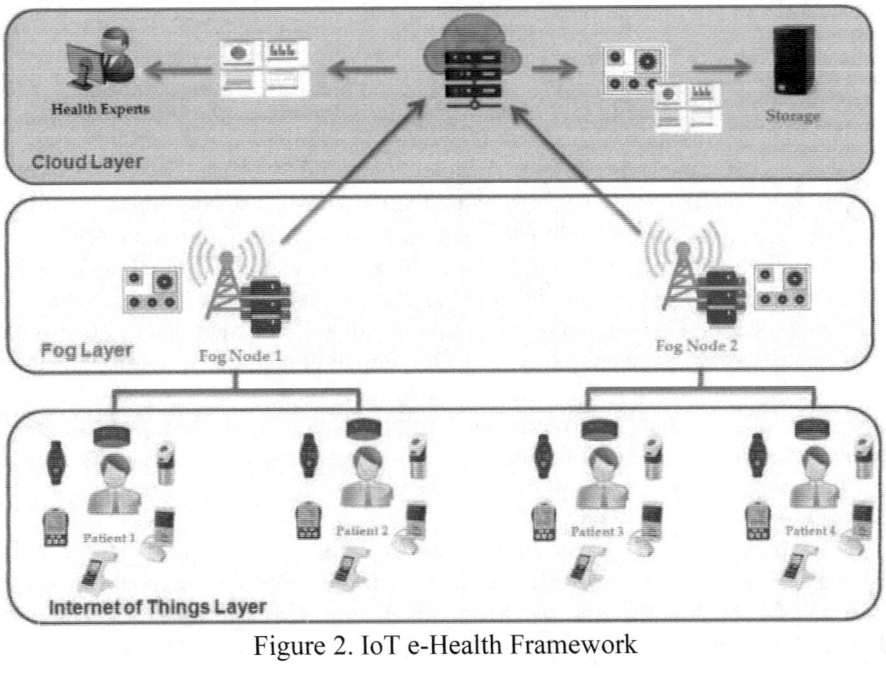

Figure 2. IoT e-Health Framework

Body Sensor Network (BSN) or Body Area Network (BAN) is a new generation of Wireless Sensor Network (WSN) installed on the human body [7]. BSN consists of biomedical sensor nodes that are small in size which aim to continuously monitor vital signs of the human body. BSN applications are very broad, ranging from applications in the military, entertainment, sports to health. The IEEE 802.15.6 standard categorizes BSN applications into two broad categories namely medical and non-medical applications (consumer electronics) [8]. The main characteristic of all BSN applications is that they are built to improve the quality of human life. Physically, the form of BSN implementation in the medical / health world is divided into three types of sensor nodes as follows [14]:

1) Implant node: Is a sensor node that is installed in the human body, where the installation is usually under the surface of the skin.
2) Body surface node: It is a sensor node that is mounted on the surface of the skin. This type is the type of node that will be used in this proposed research.
3) External node: It is a sensor node that does not come into direct contact with the skin / human body.

Photoplethysmography is a method for detecting changes in blood volume in capillaries using the property of reflection and light absorption and produces biomedical information that can be used such as heart rate and oxygen saturation [10]. There are many types of Photoplethysmography sensors that are easy to use such as clock and ring sensors. While the Piezoelectric Sensor has been used widely in the world as a system for monitoring the vital signs of the body [15]. One of the functions of Piezoelectric sensors is to be able to monitor the respiratory pattern in particular respiration rate [16]. Prediction and probability are approaches that are very often used in the KDD (Knowledge Discovery Databse) method which aims to identify something that has not happened before. The approach taken to monitor historical data on asthma patients in children can increase the vigilance of patients' parents towards their children. A general example of an asthma monitoring approach is discussed comprehensively in an article conducted by Sajeeda, et al about the extraction of detection and prevention of asthma attacks [17]. Another study conducted by Raji et al. Presented a monitoring system of respiratory rate with IoT so that it was able to carry out periodic monitoring of patients [18]. The role of IoT is also used to detect asthma attacks in real-time applications by notifying them when behavior and pattern changes occur outside normal conditions [19].

3. Methods and Approaches

In this paper we adopt a trust classification consisting of several metrics: trust metrics, trust sources, trust algorithms, trust architecture, and trust propagation, see Figure **Error! Reference source not found.** [11]. However, in this study the focus is on trust source and trust architecture for handling detection of asthma attacks in children at school or home.

1) Trust Source, this research is still open regarding the model of trust both direct, indirect or hybrid sources. A node from IoT to detect asthma attacks can use any trust source model.
2) Trust Architecture, the detection of asthma attacks the trust architecture model adjusts to the regulations of each region related to the distribution of health information. However, research recommends using a cloud-based model with the aim of several interested parties being able to read the results of asthma attack detection. Of course this kind of model needs multilevel authority. For example, interested parties related to asthma attacks in children are parents, doctors, hospitals / clinics, and the government. In Indonesia is the Indonesian Ministry of Health or Kementerian Kesehatan Republik Indonesia.

4. Result and Discussion

Based on the above theoretical considerations at this initial stage, we map papers that have been published until the latest developments. The basic approaches we use include the IoT Reference Model, Architecture of IoT e-Health, and the Model of Trust Classification in IoT. This is stated in TABLE III. about research maps and publications related to IoT in asthma attacks.

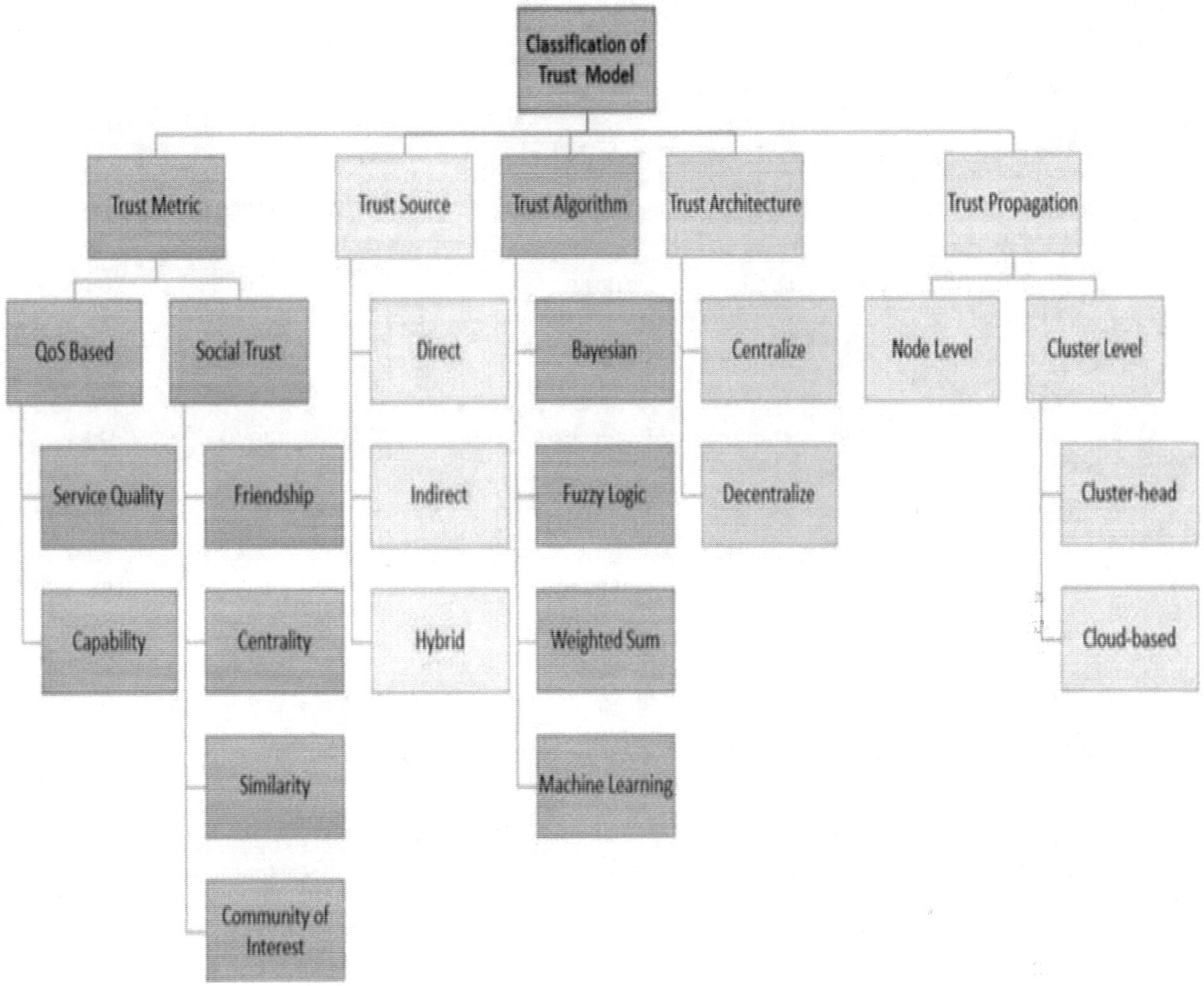

Figure 3. Model of Trust Classification in IoT

Based on the evaluation on Table 3 then we can summarize that detection devices for asthma attacks in children require a special tool to monitor asthma attacks that can provide immediate action. The criteria for the delivery of these devices must be in the form of wearables that are good in terms of design and comfortable to use for children. The device consists of a sensor that measures the pulse frequency, oxygen saturation level, and PEF Index that is connected to a direct gateway connected to the internet. This device should be in a special room or a house so that the quality of the internet network is maintained. In addition to using sensors attached to the body also need other sensors installed or using Social-IoT such as weather, air quality, temperature and air pressure. Data processing can use cloud computing technology [31] with limited authority. Users of this information include parents and medical teams that have been appointed.

So we propose the architecture of detection of asthma attacks in children that are connected with cloud computing technology that makes it easy for parents to monitor the development of a child's illness and conditions in the home and the medical team that has been assigned can provide fast and appropriate action when an asthma attack occurs in a child. Devices should be designed with the Wearable type for children's convenience that is connected directly to the IoT e-Health cloud computing system, see Figure 4.

IOP Conf. Series: Materials Science and Engineering 1232 (2022) 012004 doi:10.1088/1757-899X/1232/1/012004

Table 3. Evaluation of Related Research

Work	Trust Source	Trust Architecture	Wearable Type	Location
Filipa, 2014 [22]	Direct	Web Service IoT, Centralize	No, WSN	Home
Namazova-Baranova, 2015 [26]	Indirect	Centralize, Private Cloud	Smartphone	Hospital
Satat, 2015 [27]	Direct	Centralize	No, More WBANs	Home
Kassem, 2015 [28]	Indirect	Centralize Analytic	No, PEF, Smartphone	Any Location
Siddiquee, 2016 [19]	Indirect, Social IoT	Centralize, cloud-based	Wearable device, WSN	Any Location
Chamberlai, 2016 [29]	Indirect	Prototype using Smartphone and PC	No	Hospital, Clinic, Home
Raji, 2017 [18]	Direct	Centralize	No	Hospital
Gouma, 2017 [30]	Direct	Decentralize	Handheld	Special Room
Semernik, 2017 [20]	Direct	Stand Alone	No	Hospital
Indulakshmi, 2018 [21]	Direct	Centralize, cloud-based	WSN	Special Room, Home
Maalouf, 2018 [22]	Indirect	Centralize	WSN	Home
Silva, 2018 [23]	Indirect	Centralize	Smartphone	Any Location
Isaac, 2018 [24]	Indirect	Centralize	Smartphone	Any Location
Kadariya, 2019 [25]	Indirect	Centralize, Bot	Smartphone	Any Location

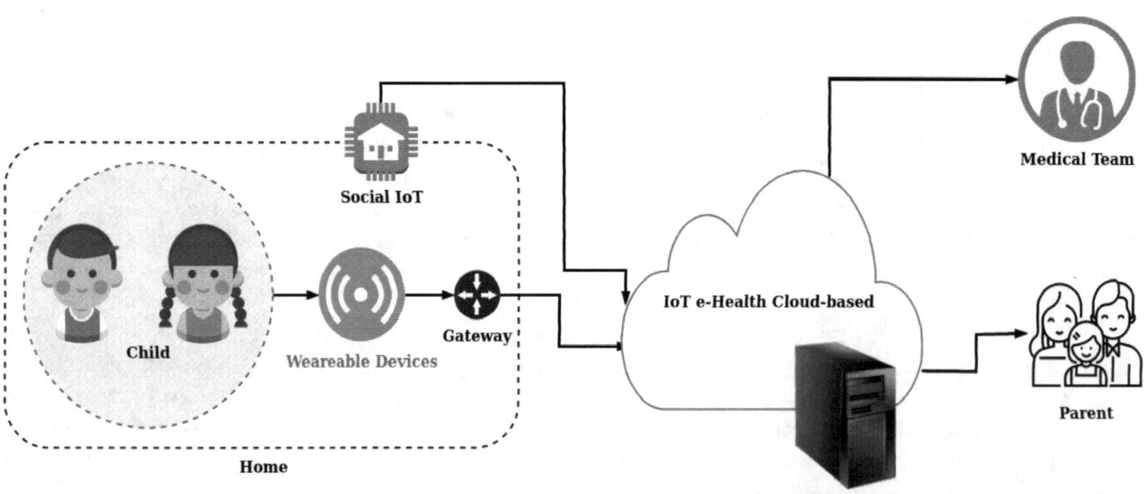

Figure 4. Proposed of Architecture Asthma Attack Detection on Child

IOP Conf. Series: Materials Science and Engineering 1232 (2022) 012004 doi:10.1088/1757-899X/1232/1/012004

5. Conclusion

IoT e-health care to anticipate sudden asthma attacks in children is designed with the type of wearable body surface node must be designed with an emphasis on the comfort of fiber does not interfere with children's activities. The IoT e-health asthma attack detection device that we designed senses the pulse frequency, oxygen, and PEF, and biases the sensor in children's play environment, especially in the weather, temperature changes, and dust density. The device and Social IoT are directly connected to IoT e-Health Cloud-based via an internet gateway. So parents and the medical team can monitor easily.

6. References

[1] Central Agency on Statistics 2018 *Statistik Indonesia 2018* (Jakarta: Badan Pusat Statistik)

[2] Global Asthma Network 2018 *Global Asthma Network 2018* (Auckland)

[3] Dharmayanti I, Hapsari D and Azhar K 2015 Asma pada anak Indonesia: Penyebab dan Pencetus *Kesmas Natl. Public Heal. J.* 9 320

[4] Ministry of Health of the Republic of Indonesia 2018 *Profil Kesehatan Indonesia 2017* ed R Kurniawan, Yudianto, B Hardhana and T Siswanti (Jakarta: Kementerian Kesehatan Republik Indonesia)

[5] Ministry of Health of the Republic of Indonesia 2018 *Hasil Utama Riset Kesehatan Dasar (RISKESDAS) 2018* (Jakarta)

[6] Adriany S 2016 *Validitas Skor Modified Pulmonary Index Score (MPIS) dan Modified Clinical Astma Evaluation Score (MCAES) dengan Peak Flow Respiratory Rate (PEFR) dalam Menentukan Derajat Keparahan Serangan Asma Akut pada Anak di Surabaya* (Universitas Airlangga)

[7] Lai X, Liu Q, Wei X, Wang W, Zhou G and Han G 2013 A Survey of Body Sensor Networks *Sensors* 13 5406–47

[8] Gravina R, Alinia P, Ghasemzadeh H and Fortino G 2017 Multi-sensor fusion in body sensor networks: State-of-the-art and research challenges *Inf. Fusion* 35 68–80

[9] Suryowinoto A, Hamid A, Fauzi Desmalasa A, Elektro T, Teknologi Industri F, Teknologi Adhi Tama Surabaya I and Arif Rahman Hakim J 2018 Deteksi Dini Penyakit Pernafasan Asma Dengan Peak Expiratory Flow Meter Berbasis Microcontroller *J. Ilm. Mikrotek* 2

[10] Lee S, Shin H and Hahm C 2016 Effective PPG sensor placement for reflected red and green light, and infrared wristband-type photoplethysmography *Int. Conf. Adv. Commun. Technol. ICACT* 2016-March 556–8

[11] Najib W, Sulistyo S and Widyawan 2019 A survey on trust in Internet of Things *The Fifth Information Systems International Conference 2019* (Elsevier B.V.) pp 1300–7

[12] Cisco Systems 2014 *The Internet of Things References Model*

[13] Maach A, Alami J E L, Houssine E and Mazoudi E 2019 A fog-driven IoT e-Health framework to monitor and control Asthma Exacerbation *2019 Int. Conf. Wirel. Networks Mob. Commun.* 0–5

[14] Abidi B, Jilbab A and Haziti M E L 2017 Wireless Sensor Networks in Biomedical: Wireless Body Area Networks *Europe and MENA Cooperation Advances in Information and Communication Technologies* Advances in Intelligent Systems and Computing vol 520, ed Á Rocha, M Serrhini and C Felgueiras (Cham: Springer International Publishing) pp 321–9

[15] Klap T and Shinar Z 2013 Using Piezoelectric Sensor for Continuous-Contact-Free Monitoring of Heart and Respiration Rates in Real-Life Hospital Settings *Comput. Cardiol. (2010).*

[16] Mahbub I, Pullano S A, Wang H, Islam S K, Fiorillo A S, To G and Mahfouz M R 2017 A Low-Power Wireless Piezoelectric Sensor-Based Respiration Monitoring System Realized in CMOS Process *IEEE Sens. J.* 17 1858–64

[17] Shikalgar S, Marathe S, Rai P and Nadar D 2015 Rule extraction for detection and prevention of asthma attacks *2015 IEEE 3rd International Conference on MOOCs, Innovation and Technology in Education (MITE)* (IEEE) pp 437–40

[18] Raji A, Kanchana Devi P, Golda Jeyaseeli P and Balaganesh N 2017 Respiratory monitoring system for asthma patients based on IoT *Proc. 2016 Online Int. Conf. Green Eng. Technol. IC-GET 2016* 1–6

[19] Siddiquee J, Roy A, Datta A, Sarkar P, Saha S and Biswas S S 2016 Smart asthma attack prediction system using Internet of Things *7th IEEE Annu. Inf. Technol. Electron. Mob. Commun. Conf. IEEE IEMCON 2016* 1–4

[20] Semernik I V., Dem'yanenko A V., Topalov F S and Nevstruev Y V. 2019 Prospects for designing a portable system for monitoring of the patient's condition with bronchial asthma *Proc. 2019 IEEE Conf. Russ. Young Res. Electr. Electron. Eng. ElConRus 2019* 1323–6

[21] Indulakshmi S, Adithya M, Anirudh A R and Jawahar A 2018 Design and Development of Prototype Model for Asthma Trigger Detection *2018 Int. Conf. Wirel. Commun. Signal Process. Networking, WiSPNET 2018* 1–5

[22] Filipa Teixeira A and Postolache O 2014 Wireless sensor network and web based information system for asthma trigger factors monitoring *Conf. Rec. - IEEE Instrum. Meas. Technol. Conf.* 1388–93

[23] Silva T A, Costa M G F, Stelmach R, Bley P K, Gutierrez M A and Filho C F F C 2018 Development of a system mobile-based to assist asthma self-management *2018 3rd Biennial South African Biomedical Engineering Conference (SAIBMEC)* (IEEE) pp 1–4

[24] Isaac N, Sampath N and Gay V 2018 Modernising Asthma Management: Personalised Asthma Action Plans Using a Smartphone Application *Int. Symp. Med. Inf. Commun. Technol. ISMICT* 2018-March 1–5

[25] Kadariya D, Venkataramanan R, Yip H Y, Kalra M, Thirunarayanan K and Sheth A 2019 KBot: Knowledge-enabled personalized chatbot for asthma self-management *Proc. - 2019 IEEE Int. Conf. Smart Comput. SMARTCOMP 2019* 138–43

[26] Namazova-Baranova L S, Molodchenkov A I, Vishneva E A, Antonova E V. and Smirnov V I 2015 Remote monitoring of children with asthma, being treated in multidisciplinary hospital *Proc. - 2015 Int. Conf. Biomed. Eng. Comput. Technol. Sib. 2015* 7–12

[27] Satat G, Ramchander K and Raskar R 2016 Identi-wheez - A device for in-home diagnosis of asthma *Proc. Annu. Int. Conf. IEEE Eng. Med. Biol. Soc. EMBS* 2016-Octob 4375–8

[28] Kassem A, Hamad M and El Moucary C 2015 A smart spirometry device for asthma diagnosis *Proc. Annu. Int. Conf. IEEE Eng. Med. Biol. Soc. EMBS* 2015-Novem 1629–32

[29] Chamberlain D B, Kodgule R and Fletcher R R 2016 A mobile platform for automated screening of asthma and chronic obstructive pulmonary disease *Proc. Annu. Int. Conf. IEEE Eng. Med. Biol. Soc. EMBS* 2016-Octob 5192–5

[30] Gouma P, Stanacevic M, Karimi Y, Huang J and Jodhani G 2017 No nanosensor and single exhale breathalyzer for asthma monitoring *ISOEN 2017 - ISOCS/IEEE Int. Symp. Olfaction Electron. Nose, Proc.* 2017-May 1–3

[31] Mell P and Grance T 2011 *The NIST Definition of Cloud Computing*

IOP Conf. Series: Materials Science and Engineering 1232 (2022) 012005 doi:10.1088/1757-899X/1232/1/012005

Production of Gummy Confection using Seeds Gum from *Durio zibethinus* Seeds

K Y Kai[1], S I Zubairi[2]*, I H Shaiful Hisham[1], M Y Ying[1], G V Lynn1, N N Lani[1], F F Fazial[2] and Z Nurzahim[2]

[1]Kolej PERMATApintar™, Pusat PERMATApintar™ Negara,
Universiti Kebangsaan Malaysia, 43600 UKM Bangi, Selangor, Malaysia.
[2]Department of Food Science , Faculty of Science & Technology,
Universiti Kebangsaan Malaysia, 43600 UKM Bangi, Selangor, Malaysia.

*E-mail: saiful-z@ukm.edu.my

Abstract. Durian is the most popular seasonal fruit in Malaysia. Only one-third of durian is edible, whereas the seeds are thrown away. This waste is potential of value-added such as seed gum. In this research, durian seed gum has been used to replace commonly used mammalian gelatine as gelling agent in making gummy confection. Several formulations consisting different compositions of gelling agent durian seed gum: fish gelatine (100:0; 75:25; 50:50; 25:75; 0:100) have been used. Experimental design was divided into two stages, preliminary and the main research. Preliminary research was conducted to produce gum extract from durian seed using different solvents. The main research was to determine the best formulation on making of gummy confection with the presence of gelatine, sucrose, glucose syrup, ascorbic acid, and pineapple juice as gel stabilizer and flavour enhancer. Selected formulation (durian seed gum: fish gelatine; 75:25) based on texture analysis and sensory test were being compared to commercial gummy confection in the market. Again, prepared gummy confection using composite durian seed gum and gelatine has better properties in terms of hardness, texture, gumminess, and overall sensory acceptance compared to individual durian seed gum and individual gelatine confection. Hence, prepared gummy candy showed good potential in the development of various types of gummy confections using waste materials, preserving nutritional values of durian origin, as well as rejecting the halal issues and bovine spongiform encephalopathy (BSE) disease regarding mammalian gelatine.

1. Introduction

Durian (*Durio zibethinus* L.; Family *Bombacaceae*) are native to Malaysia and the Southeast Asian countries. It is nominated by the locals as 'king of fruits' in Malaysia because it is the most highly prized fruit in Malaysia. The unique characteristics of this food is it has high volatile sulphur compounds. Only flesh part (also called aril) of the durian fruits is considered edible and sometimes fermented with salt. Its seed presents in light brownish color and has chestnuts like structure with length and diameter of 2-6cm, and 2-3 cm respectively. [1].

Nowadays, we can notice that most of the Malaysian likes to eat durian but they left behind a lot of durian seeds after eating the durians. Only one-third of the durian is edible, whereas the seeds (20-25%) are mostly thrown away after the consumption. Therefore, this crop waste can be significant potential source of raw material useful for the development of value-added products such as seed gum, flour and so on [2]. Natural plant gums are usually safe for oral consumption and are preferred over analogous synthetic gums due to their safety (non-toxic), low cost and availability [3]. Actually, we can utilize the durian seeds to produce something useful because of its viscid texture and it contains a lot of nutrients. Examples of nutrients that we can find in durian seeds are fibre, zinc, vitamin C, vitamin B6, manganese, potassium, magnesium and phosphorus. Based on these nutrients, durian seeds can provide a lot of benefits such as regulating blood sugar levels, promotes healthy skin, fights depression, avoids constipation and maintains healthy bones [4]. It has been proven that durian seed gum has the ability to stabilize oil in water emulsions. Durian seeds are not advised to be eaten raw because it may be poisonous, but they still can be eaten and they are very palatable when they're boiled, roasted, or fried in coconut oil. After the seeds are boiled, fried or roasted, they will have a texture that is similar to yam, but stickier [5, 6].

Gummy confectionery, or better known as gummy candy are broad category of gelatin-based, chewable sweets. Among protein-based gelling agents, mammalian gelatin has been used in manufacturing jelly confectionery due to its diverse functional properties, especially its textural, gel-forming, foam-stabilizing and emulsifying properties [7]. However, mammalian gelatin obtained from bovine or porcine sources is undesirable to vegetarian population and certain ethnic groups [8].

Fish gelatin has been used as an alternative to mammalian gelatin in recent years but its utilization is still limited due to limited access to fish skins among filleting company. Apart from this, durian seed gum has specific structure which allow intermolecular bonding with other substance due to its great compatibility for various functionalities [9]. Hence, durian seed gums could be used as the new alternative to replace gelatin in the preparation of gummy confections as the gelling agent and could be potentially marketed in the future.

Bio-polymeric materials such as seed gum (SG) can serve as good natural sources for gummy confection forming material as alternatives to the present commercial mammalian gelatin. These materials have good gel forming properties, acceptable to all religions and vegetarians, and free from viruses and prions contamination like bovine spongiform encephalopathy (BSE) and foot-and-mouth disease (FMD). Apart from this, durian seed gum has specific structure which allow intermolecular bonding with other substance due to its great compatibility for various functionalities [9]. Due to great potential of biocompatibility of both biopolymers, durian seed gum-fish gelatin (SG-FG) mixtures could be good alternative material in the preparation of gummy candy confection.

The purpose of this research is to investigate that whether the durian seeds are suitable to produce gummy confection instead of using gelatine or chemical substances such as xanthan and carrageenan as a thickening agent in the making of gummy confection. In any commercial scale of fruit processing industry, durian seed gum can be used as a thickening agent [2]. The mineral content of durian seed gum was comparable to commercial gum except for zinc content which was quite high in durian seed gum. In fact, there are low amount of the protein fraction (3.2–3.9%) presence in the chemical structure of the biopolymer from durian seed [10]. The most abundant and predominant chemical composition and molecular structure of durian seed gum are carbohydrate, fatty acid and amino acids [2]. Since, durian seed is enriched with lots of nutrient, it will be a productive option to produce gummy confection that will not bring harm to our body health because durian seeds are a waste product and completely discarded after consuming the edible flesh.

Gelatine is the main material that gives the gummy confection its characteristic properties such as good flavor release, memory and a clear, translucent appearance [11]. In gelatine and high-methoxyl pectin, the hardness, gumminess and Young's moduli of gummy confection decreased significantly with increasing water content and red pitaya fruit puree (RPFP; *Hylocereus polyrhizus*), implying a marked decrease in rigidity and formation of flexible polymer gel networks [12]. The results from texture profile analysis tests showed that in the formulation of gummy confections, the addition of acid modified corn starch to gelatine gels in suitable proportions can be a feasible alternative [13]. In the invention of gelatine-free gummy confection using gellan gum and carrageenan, the best result produced is using 2.5% carrageenan and 0.25% gellan gum E418 [14]. There are two types of gelatin. The primary sources of gelatine are from bovine animals and pigs which causes the religious ramifications arise. Other sources of gelatine are made from beef, fish or birds and this have been indicated in the literature as alternative, small volume sources of gelatine. However, these sources of gelatine providing the possible link to bovine spongiform encephalopathy, BSE, or "Mad Cow Disease" in the United Kingdom [11]. In the gummy confectionery product, mostly contain oxidized starch, optionally in combination with a gum such as gum arabic, as a replacement for at least a part and preferably all of the gelatin in the product [7].

2. Methodology

Experimental design was divided into two stages, preliminary and the main research. Preliminary research was conducted to produce gum extract from durian seed using different solvents. The main research was to determine the best formulation on making of gummy confection with the presence of gelatine, sucrose, glucose syrup, ascorbic acid, distilled water, and pineapple juice as gel stabiliser and flavour enhancer.

2.1. Extraction of Durian Seed Gum

Sample durian (*Durio zibethinus*) was collected from the Malaysian local market. Ripened and free of visual defects durian fruits were selected. The fruits were then de-husked, by cutting along the suture on the back of the durian. The durian flash was removed and the seeds was collected, cleaned and rinsed with deionized water. The seed was partially dried by the air circulation at the room temperature for one night. The skin of the seeds was peeled and the durian seeds were chopped into small sizes and dried in a cabinet dryer for 24 - 48 hours (50 °C). The seeds were mixed and sieve with 80 mesh sieve to make seed flour. The durian seed flour was extracted by using demineralised water for 1 hour and the extract were centrifuged for 10 minutes. The sediment was mixed with 96% ethanol (2 times of the supernatant) and dried for 12 hours. The extraction was being blended and sieved using a 120 mesh sieve. Preparation flow diagram of gum durian seed can be seen as Figure 1, while Figure 2 shows the final product of dried durian seed gum. The viscosity profiles was analyzed using rheometer anton paar at 25°C.

Durian Seed
↓
Washing
↓
Peeling
↓
Cutting
↓
Drying in *cabinet dryer* for 24-48 hrs (50°C)
↓
Mixing
↓
Durian Seed Flour (shieving 80 mesh)
↓
Durian Seed Flour were extracted by using demineralized water
(w/s ratio = 35.5 : 1 w/w; pH 11.9; Temp 85 °C) for 1 hour
↓
Extract (slurry) were centrifuged for 10 mins, 1200 rpm
Supernatant Sediment
↓
Mixed with 96% ethanol (food grade) 2 times supernatant volume
↓
Gum extract were dried for 12 hours
↓
The product was blended and sieved at 120 mesh
↓
Gum Extract Durian Seed Flour

Figure 1. Flow diagram of making gum extract from durian seed [5].

Figure 2. Dried durian seed gum

2.2. Preparation of Gummy Confection

Gummy confections were prepared as shown in Figure 3 using durian seed gums and fish gelatin as gelling agents. Briefly, the durian seed gum and fish gelatin were solubilized in distilled water at 60 °C for 15 minutes at various weight percentage ratios of durian seed gum to fish gelatin (100:0; 75:25; 50:50; 25:75; 0:100). Subsequently, sucrose (36.6%, w/w) was dissolved in the mixture, before adding glucose syrup (24.4%, w/w) and ascorbic acid (0.05%, w/w). Pineapple juice (3.9%, w/w) was also added as gel stabilizer and flavor enhancer in the gummy confections. Prior to molding, the mixtures were heated further until the resultant mixture reached final soluble solid content. The final mixture was then allowed to cool and set. The gummy confections prepared from gelatin incorporated with durian seed flour was referred to as Gelatin Gummy Confection (GGC). GGC samples were kept at room temperature (25 °C) until further analysis. The best formulation of confection produced was compared to the commercial gummy candy in the market using the same parameters. Then, the products were tasted by the participants.

Gum extract durian seed flour + Fish gelatin
(solubilised in distilled water for 15 min, at 60°C)

↓

Add on sucrose (36.6%, w/w)

↓

Add on glucose syrup (24.4%, w/w) and
ascorbic acid (0.05%, w/w)

↓

Pineapple juice (3.9%, w/w) was also added
(gel stabiliser and flavour enhancer)

↓

Final soluble solid content

↓

Moulding

↓

Let cool and set

Figure 3. Flow diagram of preparation of gummy confections incorporated with durian seed flour.

2.3. Formulations for Gummy Candy Confectioneries
Each formulation in Table 1 were prepared in triplicates ($n = 3$), total ($n = 15$).

Table 1. Formulation for gummy candy confection

Formulation	Sample	Ratio of gelling agent (total gelling agent = 25% w/w total mixtures)		Additives, stabilizers, and water (75% w/w total mixtures)	Total mixtures (%)
		Durian seed gum (%)	Fish gelatin (%)		
1	SG100	100	0	Sucrose (36.6%) Glucose syrup (24.4%) Ascorbic acid (0.05%) Pineapple juice (3.9%) Water (10.05%)	100
2	SG75FG25	75	25		
3	SG50FG50	50	50		
4	SG25FG75	25	75		
5	FG100	0	100		
6	Commercial	-	-	-	-

Figure 4. Gummy confection product with different ratio of seed gum: gelatin.

2.3.1. Hardness of Gummy Candy
The hardness was determined using Shimadzu Texture Analyzer. A puncture test method with needle probe was used to determine the sample hardness. The peak force obtained, in newton was taken to be the result from the test.

2.3.2. Color of Gummy Candy
The color of extracted powder was examined using chroma meter (Minolta, Japan). The color was evaluated by means of CIE Lab color components (L*; lightness, a*; redness and b*; yellowness) [15]

2.3.3. Preference test of Gummy Candy
The sensory test was participated by 30 participants. The attributes tested was taste, hardness, aroma, gumminess and potential to buy. Each participant will choose the most preferred attributes among the gummy candy samples.

3. Results and Discussion
3.1 Viscosity of durian seed gum extract
The viscosity value for the durian seed gum extract is 0.1112 Pa.s (Pascal-second), which is about100 times the viscosity of water. This was due to the high molecular weight of seed gum. The long polymeric chains linked to each other caused strong interaction between the molecules, thus prevent the liquid from easily flow [16]. This shows that the seed gum solution viscosity has high gel strength and suitable for candy making. It able to hydrate rapidly in cold water systems and forms a viscous colloidal dispersion when completely hydrated.

3.2 Hardness of Candy Sample

Hardness properties of composite gelling agent was as in Table 2. All samples except SG50F50 and SG75FG25 shows significance difference (P<0.05) to each other. The composite SG25FG75 showed highest hardness values, followed by SG50FG50, commercial, SG75FG25, FG100 and SG100 which is 4.72 ± 0.01, 4.71 ± 0.01, 2.73 ± 0.05, 2.28 ± 0.08, 1.30 ± 0.12 and 0.59 ± 0.05 respectively. It was observed that small durian seed gum (SG) amount in formulation could improve the hardness of gummy candy. Improved in hardness properties among composite gelling agent was due to protein-polysaccharide interaction between cationic domain of fish gelatine and anionic domain of durian seed gum, thus resulting in stronger synergistic network [17];[18].

Table 2. Hardness of samples

Sample	Hardness (g)	Newton unit
SG100	60.00 ± 4.72^{e}	0.59 ± 0.05^{e}
SG75FG25	232.21 ± 8.53^{b}	2.28 ± 0.08^{b}
SG50FG50	480.54 ± 0.67^{a}	4.71 ± 0.01^{a}
SG25FG75	481.31 ± 0.92^{a}	4.72 ± 0.01^{a}
FG100	132.14 ± 11.82^{d}	1.30 ± 0.12^{d}
Commercial	278.58 ± 4.78^{b}	2.73 ± 0.05^{b}

3.2. Colour of Candy Sample

Colour properties of produced gummy candy were presented in Table 3. In terms of lightness (L*), all samples shows significance difference (P<0.05) with commercial samples. FG100 shows the highest lightness, followed by SG25FG75, SG50FG50, SG75FG25 and SG100. This shows that the lightness of gummy candy increase as the durian seed gum concentration decreases. All samples was also lighter in respect to commercial sample.

The gummy candy produced using high seed gum is significantly difference (p<0.05) with the one with low concentration of seed gum. The sample SG75FG25 has the highest value in red/green coordinate (a*) at (Δ5.03 ± 0.05), however is significantly less red than the commercial sample (21.18 ± 1.45). It is observed that with the increase of durian seed gum concentration would produce redder candy. In contrast, the control gummy candy from fish gelatine (FG100) has lowest red/green coordinate (a*) at (Δ1.02 ± 0.05).

Meanwhile, for yellow/blue coordinate (b*), there is also a significance difference (p<0.05) between samples of low seed gum concentration and high seed gum concentration. The gummy candy from composite SG25FG75 has the highest and higher yellowness colour (Δ12.44 ± 0.05) than SG 100 and SG75FG25 with the value of 8.45 ± 0.7 and 9.71 ± 0.19 respectively. It was observed that low seed gum concentration increases the yellowness in the colour of the gummy candy produced.

Table 3. Colour profiles samples

Sample	Colour		
	L*	a*	b*
SG100	30.63 ± 0.92^{c}	4.85 ± 1.00^{b}	8.45 ± 0.70^{b}
SG75FG25	37.79 ± 0.69^{b}	5.04 ± 0.22^{b}	9.71 ± 0.19^{b}
SG50FG50	50.84 ± 0.43^{a}	4.31 ± 0.28^{bc}	11.97 ± 0.49^{a}
SG25FG75	51.22 ± 0.34^{a}	2.52 ± 0.07^{cd}	12.44 ± 0.30^{a}
FG100	51.86 ± 1.41^{a}	1.02 ± 0.16^{d}	11.64 ± 0.44^{a}
Commercial	28.44 ± 0.22^{d}	21.18 ± 1.45^{a}	9.19 ± 0.96^{b}

3.3 Preference test of Candy Sample

Table 4 shows sensory test results for all sample. Sample SG75FG25 (69 total scores) shows the most preferred taste, aroma, hardness, gumminess and potential to buy which is 12, 10, 16, 14 and 17 scores respectively. This shows that the most suitable formulation for the *Durio zibethinus* seed gum candy is SG75FG25.

Table 4. Sensory test of seed gum samples

Candy Rating/Sample	SG100	SG75FG25	SG50FG50	SG25FG75	FG100
Taste	4	12	9	3	2
Aroma	3	10	6	2	9
Hardness	2	16	7	3	2
Gumminess	3	14	8	5	0
Potential to buy	0	17	11	1	1
Total	**12**	**69**	**41**	**14**	**14**

4. Conclusions

As the research has demonstrated, durian seed gums are a good gelling agent that can replace gelatin in the making of gummy confection. Overall, the gummy candy with the formulation of 75% durian seed gum and 25% gelatin has the closest similarity with the commercial gummy confection. The amount of gelatin in the commercial gummy confection is higher than the gummy confection produced from this research. For future study, the usage of the other parts of the durian seeds can be studied and be applied to the daily used products. There might be other undiscovered usage of the sediments and more research can be done on this.

Acknowledgments

Authors wishing to acknowledge assistance or encouragement from Universiti Kebangsaan Malaysia (UKM) for the financial support (GUP-2018-057) colleagues and special work by technical staff at Department of Food Sciences, Faculty of Science and Technology, UKM Bangi for allowing this study to be successfully carried out.

References

[1] Husin N A, Rahman S, Karunakaran S & Bhore S J 2018 A review on the nutritional, medicinal, molecular and genome attributes of Durian (Durio zibethinus L.), the King of fruits in Malaysia. *Bioinformation* **14(6)** 265-270

[2] Amid B T, Mirhosseini H & Kostadinovic S 2012 Chemical composition and molecular structure of polysaccharide-protein biopolymer from *Durio zibethinus seed*: extraction and purification process. *Chemistry Central Journal* **6** 117

[3] Rana V, Rai P, Tiwary A K, Singh R S, Kennedy J F, Knill C J 2011 Modified gums: approaches and applications in drug delivery. *Carbohydr Polym.* **83** 1031-1047.

[4] Anon. Durians facts and health benefits https://www.healthbenefitstimes.com/durians/ [19 March 2019]

[5] Mirhosseini H, Amid B.T., Cheong K W 2013 Effect of different drying methods on chemical and molecular structure of heteropolysaccharide–protein gum from durian seed. *Food Hydrocolloids* **31 (2)** 210-219.

[6] Chung F 2011 Durian Information - A Durian Blog Devoted To "The King of Fruits". http://durianinfo.blogspot.com/ [19 March 2019]

[7] Louise B, Edward G J, Francesco S & Frank S 2006 Chewy confectionery product. *European.*

[8] Sadowska M, Kolodziejska I & Neicikowska C 2003 Isolation of collagen from the skins of Baltic cod (Gadus morhua). *Food Chemistry* **81** 257–262

[9] Arfat Y A, Benjakul S, Prodpran T, & Osako K 2014 Development and characterisation of blend films based on fish protein isolate and fish skin gelatine. *Food Hydrocolloids* **39** 58–67
[10] Amin A M, Ahmad A S, Yin Y Y, Yahya N, Ibrahim N 2006 Extraction, purification and characterization of durian (*Durio zibethinus*) seed gum. *Food hydrocolloids* 273-279
[11] Lennox S 2002 Gelatin Alternatives in Gummi Confections.
[12] Hani N M, Romli S R, Ahmad M 2014 Influences of red pitaya fruit puree and gelling agents on the physico-mechanical properties and quality changes of gummy confections. DOI: 10.1111/ijfs.12638
[13] Marfil P H M, Anhê A C B M. & Telis V R N 2012 *Food Biophysics* **7** 236 Doi:10.1007/s11483-012-9262-3
[14] Grazela A J, Morrison N A 2003 Gelatin-free gummy confection using gellan gum and carrageenan.
[15] Savedboworn W, Phuditcharnchnakun S & Khuntaweetap T 2015 Development of antioxidant gummy jelly candy supplemented with psidium guajava leaf extract. *Int J Appl Sci Technol,* **8(2)**145-151.
[16] Ahmed E M 2015 Hydrogel: Preparation, characterization, and applications: A review. Journal of Advanced Research **6 (2)** 105-121.
[17] De Kruif C, & Tuinier R 2001 Polysaccharide protein interactions. *Food Hydrocolloids* **15 (4-6)** 555–563.
[18] Pranoto Y, Lee C M, & Park H J 2007 Characterizations of fish gelatine films added with gellan and κ-carrageenan. *LWT - Food Science and Technology* **40 (5)** 766–774.

Structural Equation Model (SEM) Correlation Between *Work Breakdown Structure* (WBS), Work Method and Risk Towards Cost Of Safety On Low-Cost Apartments Project

R Hadwiansyah and Y Latief

Faculty of Civil Engineering, University of Indonesia, Indonesia
Email: riezka.hadwiansyah@gmail.com, yusuflatief73@gmail.com

Abstract. Cost of safety is a fundamental aspect that is inevitable to be considered during the construction cost planning phase. The cost of safety estimation is determined based on several factors, including WBS (work breakdown structure), work method, risk and other contributing factors. This estimation is critical in construction, specifically for high rise building constructions. In a case where there is a lack of a suitable methodology in estimating the cost of safety, may incur inaccuracy and uncertainty in determining the actual cost of safety. Besides the mentioned factors, this study will use Structural Equation Model – Partial Least Square (SEM-PLS) in determining the construct model cost of safety, for the case of the build of low-cost apartment building (Rusunawa).

Keywords: SEM, Cost of safety, Apartment, WBS, Risk, Work Method

1. Introduction

Rusunawa, also known as a Low-cost apartment, has been a great solution to solved the overpopulated city, providing housing for low-income or non-permanent jobs with contract-based residents. It also becomes a solution for new families who are not eligible for credit housing (KPR) [1]. Along with the construction of low-cost apartments, project developments achieve positive performance and have its downside. Based on the BPJS statistics data, there are approximately 110,285 workplace accidents in construction sites in 2015, and 101,637 accidents in 2016. It was observed that there has been an increase in the following year, with 123,041 recorded accidents, and 174,105 accidents in 2019. The increase of accidents up to 40% is caused by the increase of projects around Indonesia, based on Presidential Regulation No. 3 of 2016 on Acceleration of Strategic National Project Implementation. The number of accidents in construction project sites will potentially increase in 2019, due to the increasing number of projects compared to previous years [2].

Developing construction project plans requires adequate planning from the design plan phase, construction, and the final handover. According to Ervianto [3], a construction project is a series of activities that are only carried out once and are generally short-term. In addition, construction projects also have unique characteristics, require resources (manpower, materials, machines, money, methods), and require organization. A study conducted by the Construction Industry Institute in 2012 revealed poor performance in the industrial project sector, where nearly 70% of projects exceeded the 10% variation of the expected cost and schedule [4] . An important part of the project cost element is OSH costs. OSH cost is an expense or cost for accident prevention measures undertaken by contractors [5]. According to Ministry Regulation of Public Work No. 05, 2014, the cost of OSH Construction in the Field of Public Works is the cost required to implement Safety and Health Management System in every construction work that must be calculated and allocated by the Service Provider and the Service User. The cost of administering OSH in Indonesia is regulated in Circular Letter No. 66 of 2015 concerning the Cost of Implementing an Occupational Safety and Health Management System in the Field of Public Work [6].

WBS will help analyze the components of each job in more detail and accurately. [7] WBS has 4 important objectives, such: 1) As a planning and design tool that can help project managers and teams to identify and manage projects effectively. WBS is a tool for designing structures that represent relationships between project unit clearly; 2) As a planning tool to show the completeness of the work packages in the project and detailed instructions for completing the project in each work unit; 3) Being a tool to determine the job *milepost,* project status reports to senior managers and customers, and the estimated results of each work WBS can prevent negligence of less / wrong project end results and help managers monitor project objectives and clarify tasks estimate workload; and 4) As a tool to accurately estimate time, cost, and resources.

The characteristics of WBS depend on the nature of the project and how the project manager wishes to plan and manage the project [8]. Decomposition of the WBS-based functions is being utilized to easily manage the deliverables. In planning, WBS has 5 main indicators in the management process [9]: Work package, Job package description, Person in charge, Reference document, and Required resources.

Referring to the Regulation of the Minister of Public Works and Public Housing of the Republic of Indonesia Number 21 / PRT / M / 2019, the method is a series of construction implementation activities that follow procedures and have been designed in accordance with knowledge or standards that have been tested. For construction works that use labor-intensive methods or use a lot of labor but use little machine tools, the need for Construction Safety Personnel is determined by the Construction Safety Risk assessment. This describes the components of the work method into 3 (three) aspects: scope of work, job description, and stages of work [10].

Assessment of Risk Safety Construction is the calculation of the amount of potential is based on the possibility of the existence of events that have an impact on the loss on the construction, the soul of man, the safety of the public, and the environment that may arise from certain sources of danger, occurs in Construction Work by calculating the frequency value and the severity value. Thus, the level of risk on project construction is set in the Risk Safety Construction and is categorized into 3 (three) levels: low risk, medium risk, and high risk [11].

On the Regulation of the Minister of Public Works and Housing of the People of the Republic of Indonesia number 1 Year 2020 About the Standards and Guidelines for Procurement of Works Construction Integrated Design and Build Through the Provider has set some conditions on the cost of OSH are as follows: 1) Cost of Occupational Health and Safety (OSH) Construction of a general nature in accordance with the Regulation of the Minister of Public Works No. 5 of 2014 on Guidelines for Safety and Health Management System (SMOSH) The work of the Sector Construction Works included into the cost of the general cost of OSH; and 2) Calculation of costs for the purposes of OSH which are general as Personal Protective Equipment: helmets, vests, boots, mask, rain jacket, hat, safety glove, eye protectors and others already calculated in the cost of general overhead

2. Method

2.1. Research design

In this research we used some approach to answer the research question. First, the Survey method is used to obtain the research data derived from sources such as the conduct of questionnaires, interviews, and tests aimed to observe and record it systematically on phenomena that were investigated [12]. The method is used to obtain the data used in identifying the plan of salvation that is used and validation of the results of the analysis of the variables that affect the cost of OSH. Second, Case studies are intended to determine the existing condition of the object of research, so that information on the factors that influence OHS costs can be known. The last, SEM method is a statistical technique that is able to analyze patterns of relationships. This method aims to analyze the pattern of relationships between latent constructs and their indicators, latent constructs with each other, and direct measurement errors. The SEM method also allows direct analysis of the dependent and independent variables. The overall research design is shown in Figure 1.

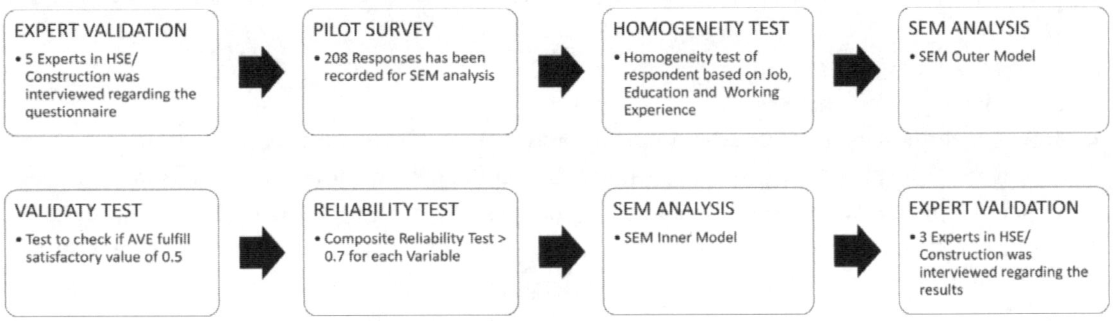

Figure 1 Research Flow Diagram

A research variable is an attribute or nature or value of people, objects, or activities that have certain variations that are determined by researchers to be studied and conclusions can be drawn. The variables in this study are: 1) WBS (X1); 2) Working Method (X2); 3) Risk (X3); 4) General cost of OSH (Y1); and 5) Specific costs of OSH (Y2).

2.2. Instruments

The instrument is a tool aids selected and used by researchers in its activities to collect the data, so that the activities are becoming systematic and can be easy. Instruments of research are referred to also as a means of measuring in a study [14] The collection of data aimed to test the hypothesis that has been formulated. The instrument of research uses SEM (Structural Equation Model) and the questionnaire as a tool to help research. This research uses a questionnaire as a material or tool for data collection. As for stage questionnaires, d nature of research is there are several, namely:

1). Stage I Questionnaire

Surveying experts with the help of instruments questionnaire early (form validation experts) regarding:

a. WBS, work methods, and risks that can affect OHS costs. These factors were obtained from literature studies. At the stage early, factors of each variable results of the study of literature is verified and clarified by experts.

b. The level of impact is indicated by the indicators that exist on the project.

2). Phase II Questionnaire

Once the process of verification, clarification, and validation of experts, deployment of instruments questionnaire phase II, a pilot survey. Questionnaires were given to prospective respondents on stage III questionnaires. On stage this, will obtain the results if the variables in the questionnaire is easy to understand.

3). Questionnaire Phase III

The submission of the stage III questionnaire through a survey of respondents. The questionnaire given to respondents to choose the level of influence of the variables are generated on the instrument questionnaire phase III. The survey questionnaire phase III given to respondents that the parties executing a job which involved directly in technical in the work of construction.

4). SEM-PLS

The next stage is the collection of primary data, then the data is obtained and analyzed using statistical software, Statistical Product and Service Solutions (SPSS), and Structural Equation Modeling (SEM-PLS) software. The SEM method is a statistical technique that can analyze patterns of relationships. This method aims to analyze the pattern of relationships between latent constructs and their indicators, latent constructs with each other, and direct measurement errors.

2.3. Data Analysis

2.3.1. Homogeneity Analysis

Homogeneity test is a test of whether or not the variances of two or more distributions are equal. Homogeneity test is performed to determine whether the data in variables X and Y are homogeneous or not. Homogeneity testing is then performed using the help of the SPSS 22 program. The assumptions used are likely to be the same, so the significance value used is 0.05 where if the significance value exceeds 0.05, it can be said that the variance of two or more groups is the same.

The data test technique used Kruskall Wallis H. One-Way Variety Method Kruskall Wallis H test is a rank-based nonparametric test whose purpose is to determine whether there are statistically significant differences between two or more groups of independent variables on the dependent variable on a numerical / ratio scale and ordinal scale . The test carried out was the free K sample test with the Kruskall Wallis H test with education groups, positions, and work experience in the OSH field.

2.3.2. Variable Correlation with *Smart* PLS

a. Validity Test

Validity test is required with the aim to find out the level of validity of the instrument and the survey results of respondents that have been conducted by researchers. This validity test is done by using a standardized loading factor value obtained from SEM Smart PLS analysis results. 3.3.2 Standardized loading factor is used to assess whether the extract has adequate discriminant validity or not, by comparing the correlation indicator of a construct with the correlation of the indicator with another construct. If the correlation value of a construct has a higher value than the correlation of other construct indicators, then it can be said that the contract has a higher discriminant validity. Standardized loading factor can describe the magnitude of the correlation of each indicator where the indicator loading factor is more than 0.7 can be said to be valid as an indicator of measuring the structure. [15] suggested that all items in the factor model must have a communality of more than 0.60 or a community average of 0.7 to justify the justification of factor analysis. Because indicator values above 0.7 can be said to be ideal. However, if there is an indicator loading factor value below 0.7 and above 0.5 (range between 0.5 - 0.7), the indicator can still be used. By opening the main menu on the toolbar calculate in the SmartPLS 3.3.2 program there is a PLS Algorithm option. This step is often referred to as First Order Confirmatory Factor Analysis.

b. Reliability Test

Reliability test is done by looking at the composite reliability value of the indicator block that measures the construct. The results of composite reliability will show a satisfactory value if above 0.7. A test of the adequacy of the data is needed to ensure that what has been collected and presented in the weighing report is objectively sufficient. Test the adequacy of the survey data of the results of respondents can be done using the KMO (Kaiser Meyer Olkin Measure) test obtained from the analysis using SPSS 22 application assistance in the Dimension Reduction - Factor. The data can be categorized as sufficient if the KMO value is above 0.5

c. Evaluation of the Inner Model

Evaluation of structural models (inner model) is an evaluation of structural models to predict causality relationships between latent variables. Using a bootstrapping procedure that aims to predict the relationship between latent variables used in evaluating structural models by looking at the magnitude of the percentage of variance described to observe the value of R2, to construct latent endogenous, Stone - Geisser ' s = Q2 to test predictive relevance and average variance extracted (AVE) for predictiveness by using resampling procedures . This procedure uses all original samples for resampling, which should be greater than the original sample. The minimum sample size in the number of bootstrap samples needed to reduce bias in all types of SEM estimates is 200 and is sufficient to correct the standard error estimate PLS [16]. Based on the complexity and characteristics of the model [17], the measurements are explained as follows: 1) If the model contains 5 or less constructs, where each construct is measured by more than 3 indicator items that have a large enough community (0.6 or more), then the required size is 100 to 150; 2) If there is moderate communality (0.45 - 0.55), or the model contains constructs measured by less than 3 indicator items, then the required sample size is more than 200; 3) If there is low communality or the model contains under identified constructs, then the required sample size is at least 300; and 4) If the model contains more than 6 constructs, of which are measured using less than 3 indicator items and have low communality, then the required sample size is at least 500.

From the results of the bootstrapping procedure on Smart PLS on outer loading, the relationship between the indicator and its latent variable is through the value of the T statistic. A statistical T value of more than 1.96 indicates a significant relationship between the indicator and its latent variable. The greatest statistical T value indicates the most dominant indicator in a variable.

3. Results and Discussion

Model Relationships SEM PLS by using *software Smart* PLS 3.3.2 is shown in Fugure 2, and Table 1 shows the *loading factor* of each indicator for the variable.

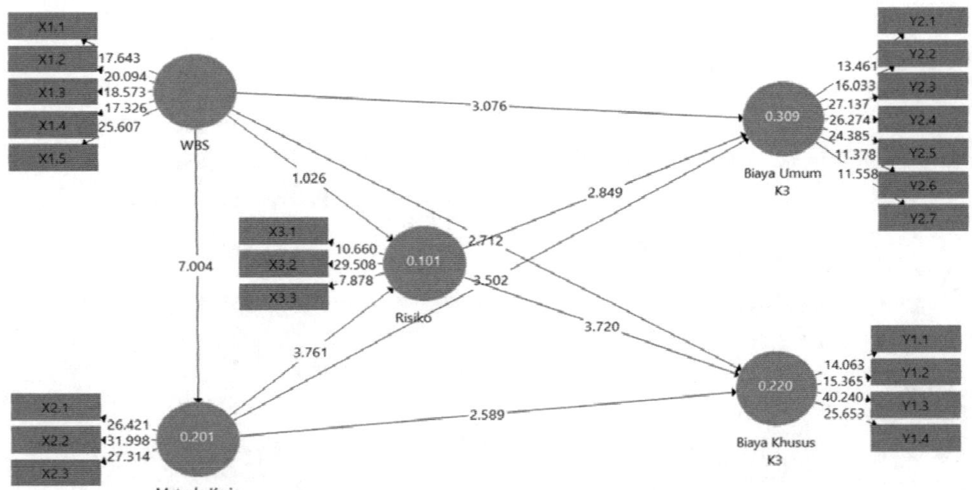

Figure 2 Model Relationships SEM PLS by using *software Smart* PLS 3.3.2

Table 1 The *loading factor* of each indicator for the variable

	WBS	Work Method	Risk	General cost of OSH	Specific cost of OSH
X1.1	0.766				
X1.2	0.786				
X1.3	0.769				
X1.4	0.780				
X1.5	0.799				
X2.1		0.852			
X2.2		0.867			
X2.3		0.843			
X3.1			0.778		
X3.2			0.935		
X3.3			0.670		
Y1.1				0.814	
Y1.2				0.821	
Y1.3				0.882	
Y1.4				0.829	
Y2.1					0.735
Y2.2					0.783
Y2.3					0.819
Y2.4					0.799
Y2.5					0.774
Y2.6					0.656
Y2.7					0.644

The loading factor value obtained in each indicator is more than 0.5 thus, it can be concluded that all indicators are valid because they meet the convergent validity value. Furthermore, the reflective indicators also need to be tested for discriminant validity by looking at the value of the square root of average variance extracted (AVE). The recommended value is above 0.5.Table 2 is the AVE value in this study:

Table 2 AVE values for each variable (Validity Test)

Variables	Average Variance Extracted (AVE)
Work Breakdown Structure	0.609
Work Method	0.730
Risk	0.643
General cost of OSH	0.558
Specific cost OSH	0.700

Table 3 Value *Composite Reliability* (Reliability Test)

Variables	Cronbach's Alpha	Composite Reliability
Work Breakdown Structure	0.840	0.886
Work Method	0.815	0.890
Risk	0.709	0.841
General cost of OSH	0.867	0.898
Specific cost OSH	0.858	0.903

Table 4 *Path Coefficient Bootstrapping* Results *(Correlation* Between Variables)

No	Correlation Between Variables	Original Sample(O)	T Statistics (|O/STDEV|)	Information
1	*Work Breakdown Structure Work Items* → Work Methods	0.048	7.004	Significant
2	*Work Breakdown Structure Work Items* → Risks	0.090	1.026	Not Significant
3	*Work Breakdown Structure Work Items* → General Costs of OSH	0.234	3.076	Significant
4	*Work Breakdown Structure Work Items* → Specific Cost of OSH	0.182	2.712	Significant
5	Work Method → Risk	0.268	3.761	Significant
6	Work Method → General Cost of OSH	0.301	3.502	Significant
7	Work Method → Specific Cost of OSH	0.192	2.589	Significant
8	Risks → General Costs of OSH	0.205	2.849	Significant
9	Risks → Specific cost OSH	0.260	3.720	Significant

IOP Conf. Series: Materials Science and Engineering 1232 (2022) 012006 doi:10.1088/1757-899X/1232/1/012006

In the path coefficient table 3 and 4, to see the significance of the influence between variables, it is necessary to look at the t- value (T-statistics), which is> 1.96. For example, Number 1, Work Breakdown Structure affecting Work Methods, has a value of 7.004 > 1. 96, it can be said that Work Breakdown Structure has a significant correlation on Work Methods.

Work Breakdown Structure is reflected by 5 (five) indicators (Table 5), the work method (X2) is reflected by 3 (three) indicators (Table 6), and Table 7 shows X3 risk are reflected by 3 (three) indicators.

Table 5. Indicators Variable X1 WBS

WBS Variable (X1)		Indicator		T Statistics (| O / STDEV |)
X1	WBS Work Package	X1.1	Work package	17.643
		X1.2	Job description	20.094
		X1.3	Person in charge	18.573
		X1.4	Reference document	17.326
		X1.5	Resource needed	25.607

The dominant indicator among the five indicators above is the Resource needed with T *Statistics* = 25.607

Table 6. Indicators Variable X2 Methods of Work

Work Method Variable (X2)		Indicator		T Statistics (| O / STDEV |)
X2	Work Method	X2.1	Scope of work	26.421
		X2.2	Job description	31.998
		X2.3	Stages of work	27.314

The dominant indicator is Job Description with T *Statistics* = 31.998.

Table 7. Variable Indicator X3 Risk

Risk Variable (X3)		Indicator		T Statistics (| O / STDEV |)
X3	Risk	X3.1	Low risk	10.660
		X3.2	Medium risk	29.508
		X3.3	High risk	7.878

The dominant indicator is Medium risk with T *Statistics* = 29.508.

4. Conclusions

Based on the tables R 2 (R- Square results bootstrapping) we have obtained value R Square of General cost of OSH 0.220 and Specific cost of OSH by 0.309. It means that the variability construct Cost of OSH can be described but not related directly in closely by the Work package, Job description, Person in charge, Reference document, Resource needed, Scope of work, Job description, Stages of work, low risk, medium risk and high risk with the value of the relationship of 22% on General of OSH and 30.9% for Specific cost of OSH. Based on the results of Smart PLS 3.3.2 with bootstrapping methods (Path Coefficient) discovering 9 the relationship between variables were observe to match the significance (T Statistics> 1.96) where 8 (eight) relationships between the variables were found significant and have mutual influence which the General cost of OSH is strong influenced by work method while Specific cost of OSH is strong influenced by the risk .

Acknowledgments
The author would like to thank the financial support provided by the University of Indonesia through PUTI Proceedings of the Fiscal Year 2020 with the contract number: NKB-1184 / UN2.RST / HKP.05.00 / 2020 managed by the Directorate of Research and Community Engagement (DRPM) University of Indonesia.

References

[1] PUPR. (2016). *Modul Pemanfaatan Rusunawa PUPR.*

[2] ISafety. (2018). ISafety Magazine. *BPJS TK*, Indonesia

[3] Ervianto, I. W. (2005). Manajemen Proyek Konstruksi Edisi Revisi, , Indonesia

[4] CII. (2012). Construction Industry Institute University of Texas, Houston. (n.d.). Quality in Construction. : https://www.construction-institute.org.

[5] Feng, Y. (2014). Exploring the interactive effects of safety investments, safety culture and project hazard on safety performance: An empirical analysis.

[6] PUPR. (2019). Pedoman Sistem Manajemen Keselamatan Konstruksi.

[7] Su, L. (2012). WBS-based Risk Identification for the Whole Process of Real Estate Project and Countermeasures.

[8] Ajizah, N. (2018). Perencanaan Sumber Daya Pada Pekerjaan Mekanikal Dan Elektrikal Bangunan Gedung Apartemen Berbasis Wbs(Work Breakdown Structure).

[9] Latief, Y. (2019). Pengembangan Kamus Work Breakdown Structure (WBS) Pada Pekerjaan Konstruksi Sturktur Bawah Jembatan Baja. Pengembangan Kamus Work Breakdown Structure (WBS) Pada Pekerjaan Konstruksi Sturktur Bawah Jembatan Baja.

[10] PUPR. (2014). Pedoman Sistem Manajemen Keselamatan Dan Kesehatan Kerja.

[11] PUPR. (2019). Pedoman Sistem Manajemen Keselamatan Konstruksi.

[12] Basrowi, &. S. (2012). Memahami Penelitian Kualitatif. Jakarta: Rineka Cipta.

[13] Basrowi, &. S. (2012). Memahami Penelitian Kualitatif. Jakarta: Rineka Cipta.

[14] Sugiyono. (2007). *Metodologi Penelitian Bisnis.* PT. Gramedia, Indonesia.

[15] MacCAllum. (1999). Sample Size in Factor Analysis: The Role of Model Error.

[16] Loehlin, J. C. (1988). *Latent variable models: An introduction to factor, path, and structural analysis (3rd ed.).*

[17] Hair, J. F. (2011). *PLS-SEM: Indeed a silver bullet.*

Risk Assessment of Electronic Archive Services using Octave Allegro Method (Case Study: SIKN JIKN)

R A Herdianto[1*], K Ramli[1], Y Suryanto[1]

[1]Department of Electrical Engineering, Universitas Indonesia, Indonesia

*Coresponding author: rizal.aditya@ui.ac.id

Abstract. The National Archives Information System (SIKN) and the National Archives Information Network (JIKN) are the government national priority programs that aimed at providing web-based electronic archive treasury information facilities to the public. However, threat and disruption of critical information asset on electronic archives can cause interruption strategic objectives goals, vision, and mission of The National Archives of Indonesia (ANRI) in organizing archives that impact the network node, and the public. Therefore this research aims to analyze and identify risks in the SIKN JIKN electronic archive service and determine the potential risk of threats to important information assets. The OCTAVE Allegro method used to measure information security on five important information assets of the SIKN JIKN electronic archive information service. The results showed 64 areas of concerns analyzed, as many as 40 had to mitigate while the remaining 24 could be deferred or postponed. The conclusion of the risk assessment results in recommendations that must take to manage and protect critical assets in electronic archive services and supporting material for policymaking.

1. Introduction

The emergence of information technology has led to changes in the transformation of archives from conventional to electronic or digital which are increasingly efficient such as reducing physical damage to the original archive, archiving data transmission through a computer network using the metadata archive exchange protocol, assessing archive retention, processing, and storing archives [1]. The National Archives of Indonesia (ANRI) is a government agency with the main task of organizing archives [2]. Archiving services are indispensable for accessing archive information for the public or users and a source of treasures for historians and researchers in health, economy, society, and culture. In the aspect of the health sector, archives have an essential role in preservation, maintenance, and services during the pandemic as well as a media source of information on handling outbreaks about Covid-19 disease for future research [3].

SIKN or The National Archives Information System and JIKN or the National Archives Information Network are a dynamic archive and static archive information service facility that aims to provide electronic records access facilities for the public. The role of SIKN comprehensively collects archived information in the form of archive description. A custody institution manages it at the provincial, district, or city level called the Network Node. JIKN is used to provide archived information for the people gathered from various network nodes [4].

IOP Conf. Series: Materials Science and Engineering 1232 (2022) 012007 doi:10.1088/1757-899X/1232/1/012007

However, threats and disruptions to information technology can interruption the organization's primary business processes in the form of cyber-attacks, loss of electricity resources, damage, or loss of archive data due to negligence by the employees themselves [5,6]. This is a disorder that often occurs, especially in government agencies [7]. The threat of risk to electronic archives is very susceptible to changes in the authentication of the archive information itself [8]. If not appropriately managed, management of risk threats will harm ANRI in archiving services that will impact archive users or the public.

Therefore need a risk assessment method to anticipate threats and disturbances to electronic archive information [9]. This study discusses the formulation of the problems regarding risk assessment of the SIKN JIKN information system. Limitations in this study include two things, namely business processes related to electronic archive information services with the scope of data, applications, IT supporting infrastructure (software and hardware), and employees.

The basis for risk assessment in this study uses the Allegro OCTAVE method to assess a large operational risk environment with better risk assessment results and focus on important information assets that are owned and have an effect on the organization. The importance of information assets for organizations how these assets are used, transferred, processed, and stored as well as how threats, vulnerabilities, and disruptions cause business process obstacles to the organization [6]. The stages of risk assessment are arranged sequentially into four phases, carried out into eight steps. The four phases include the first Establish Driver is the organization developing risk measurement criteria based on the organization's strategic goals, vision, and mission. The second, Profile Assets, is to create and determine critical information assets for the survival of an organization's business. The third, Threat Identification is developing the risk profile of information assets by identifying threats that impact information assets and threat scenarios based on three platforms (technical, physical, and people). Fourth, by identifying risks and determining mitigation approaches from the results of identification and analysis of threats or disturbances to critical information assets and developing strategic steps to overcome the risks that occur.

2. Research Methodology

The research began by conducting interviews with the Head of SJIKN Center, Head of System Development and JIKN, Archivists, and related IT Staff to explore existing problems, policies, hopes, and desires to conduct direct observations to obtain the data needed. Also, interviews were carried out in more depth to explain the importance of information technology risk assessment to the respondent. There are eight steps to the risk analysis and evaluation of the SIKN JIKN system in the risk assessment in the OCTAVE Allegro method.

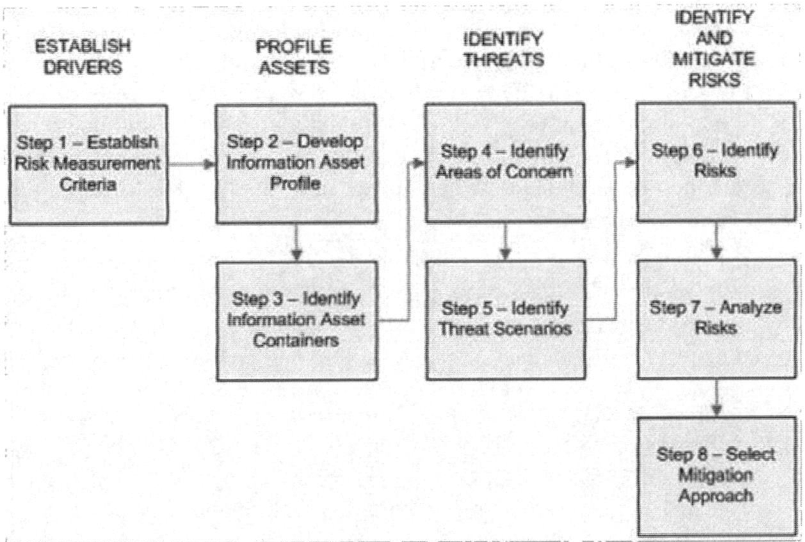

Figure 1. Steps for Risk Assessment in the OCTAVE Allegro method [6].

3. Results and Discussion

The methodology stage, the following are the results of the analysis of each phase carried out in this study.

3.1. Establish Risk Measurement Criteria

In this first stage consists of two activities in identifying the area of impact of the organization through risk measurement criteria. The first activity is to formulate and qualitatively measure the area of risk impact determined by considering the effect on the organization in the event of threats and disturbances as in table 1. Then, the second activity of the results of the impact area is determined by prioritizing the scale with the most critical value (highest score) to the least important (lowest score) as in table 2. The priority scale of the area of impact will be used in risk assessment to develop a relative risk score that can help the organization decide what mitigation measures to take.

Table 1. Example of Risk Measurement Criteria - Reputation & Customer Confidence

Impact Area	**Low**	**Medium**	**High**
Reputation	Organizational reputation has little effect; no business or small business is needed for repairs.	Organizational reputation is quite bad, and it takes effort and cost to repair.	Organizational reputation has a bad impact, and it takes effort and cost to repair.
Customer Confidence	There is user trust to access archived information access through SIKN JIKN.	There is still the confidence of some users who still access archived information through SIKN JIKN.	No user trust accesses archived information through SIKN JIKN.

Table 2. Example of Risk Measurement Criteria - Impact Areas Priority

Priority	**5**	**1**	**3**	**4**	**2**
Impact Areas	Reputation & Customer Confidence	Financial	Productivity	Safety and Health	Fines and Legal Penalties

3.2. Develop an Information Aset Profile

In the second stage, the OCTAVE Allegro method has eight activities that are described in the critical information asset form. It starts by identifying groups of information assets that are related to the organization's operational business processes. The second and third activity, consider the real impact of critical information assets that are the main objective and vision and mission of the organization in the event of threats and disturbances and then compiled into the Critical Information Asset Profile form. The fourth activity explains the reasons for selecting critical information assets in this profile. Fifth, make a general description to explain the information assets and identify who owns the information assets. The last activity identifies the potential risk impact of information assets on the most important security requirements.

Tabel 3. Example of Critical Information Asset Profile – Archival Description

Critical Aset	Archival Description
Rational for Selection	Archival description plays a role in identifying, managing, discovering or searching, explaining the records in question, the context, and the management system of the archives that created them, and supporting the exchange of archive data from different systems.
Description	Archival description is information attached to an archive to explain the context and contents that are useful in searching archive information so that it is easily identified, managed, and accessed. It has standard data elements that are used in providing widely structured or semi-structured descriptive information in archives.
Owner	Network Node

Critical Aset		Archival Description
Security Requirement	Confidentiality	Archival metadata information is vital for network centers, network nodes, and archive users. Only the Archive Manager or Network Node can authorize the archive information.
	Integrity	Archival metadata information must be correct following the results of the archive description. It can be changed by archivists and archive managers or network nodes.
	Availability	Archival metadata information in open categories must always be available to archive users.
Most Important Security Requirement		**Integrity:** Archival metadata information is vital for archive users if there is a change in the archive metadata information. It makes it difficult for archive users to search the source or archival treasures.

3.3. Identify Information Aset Containers

This stage identifies by considering three critical issues related to security and the concept of the container of information assets based on the risk environment map such as technical, physical, and people. This container describes how information assets are protected from vulnerabilities or threats to information asset containers are managed internally and externally.

Tabel 4. Example of Information Asset Risk Environment (Technical) – Archival Description

Container Description		Owner(s)
Inter	Archive metadata stored by the Network Node Admin through the SIKN application and displayed through the National Archives Information Network (JIKN) application.	Development of SJIKN
External	Network Node Admin uses the SIKN Application to store and manage archive metadata, create reporting results from report generate features, store metadata in bulk, synchronize metadata with XML standards.	Network Nodes
	Archive users search archival information based on archive metadata displayed through the JIKN application.	Archive Users

3.4. Identify Areas of Concern

In the fourth step, it starts by developing a risk profile for the SIKN JIKN information assets by conducting interviews with related parties to determine the threat component of the situation and conditions that can threaten the information assets of the archive.

Tabel 5. Example of an Area of Concern

Area of Concern
There is an interruption in the RHV server down impact on the JIKN application so that archive users cannot access the archive metadata.

3.5. Identify Threat Scenarios

In the fifth step, identify a broader threat scenario that has not been accommodated in the Area of Concern through questionnaires. Threat scenarios are arranged in more detail based on the threat of tree structure. Each scenario is identified and described based on the characteristics of the threat property.

Tabel 6. Example of Information Asset Risk Environment (People) – Incompleted of Information

Area of Concern	Threat of Properties	
Errors and incompleteness of archived metadata information cause SIKN Application report results to be inappropriate.	**1. Actors**	Admin SIKN
	2. Means	SIKN admins do not input complete mandatory elements of archive metadata
	3. Motives	Human Error (Accidental)
	4. Outcome	Destruction/Loss
	5. Security Requirements	SIKN admins ensure the mandatory input of mandatory archive metadata elements.
	6. Probability	Medium

3.6. Identify Risk

In the sixth step, determine the consequences of the threat scenarios developed and the impact on the organization or owner of critical information assets. The results of consequences are measured based on the qualitative value of the impact area and calculate the quantitative score calculated from each of the impact area values and the result of the entirety.

Tabel 7. Calculate Risk of Impact Area Scores

Impact Areas	Priority	Low (1)	Moderate (2)	High (3)
Reputation and Customer Confidence	5	5	10	15
Financial	1	1	2	3
Productivity	3	3	6	9
Safety and Health	4	4	8	12
Fines and Legal Penalties	2	2	4	6

3.7. Analyze Risk

The seventh step is about the risk analysis process by giving a quantitative value of how much influence the threat poses to the organization. The first activity starts by considering the risk measurement criteria. The second activity calculates the relative value of risks used to analyze risks and determine the best strategy to reduce the impact of the risk.

Tabel 8. Example of Properties of Threat – Corrupt Information Archival Descriptions

Area of Concern	Risk			
Archived metadata information is corrupt, or data does not appear during the process of changing the archive structure, updating the status of the publication, or the process of mass import through a browser.	Consequences	The perception of the quality of archive information for archive users gives a negative impression on the organization's services. There are additional IT Staff working hours to analyze, repair, and restore access. As a result of interference, the level of information security on legibility and access to archived metadata is low.		
	Severity	**Impact Area**	**Value**	**Score**
		Reputation & Customer Confidence	High (3)	15
		Financial	Low (1)	1
		Productivity	Moderate (2)	6
		Safety and Health	High (3)	12
		Fines and Legal Penalties	Low (1)	2
			Relative Risk Score	**36**

3.8. Select the Mitigation Approach

At this last stage, it is to sort each risk that has been identified based on the value of risk analysis in step seventh. This sorting activity is carried out to assist the organization in deciding the status of the type of mitigation approach that needs to be done at the risk area of concern.

Table 9. Relative Risk Score

	Relative Risk Matrix		
Risk Score	**30 to 40**	**16 to 29**	**0 to 15**
	Pool 1	Pool 2	Pool 3

The calculation of quantitative risk scores is the probability of the area of concern and the consequences to determine one of the status categories such as mitigation, defer, or accept in figure 2.

Tabel 10. Determine the Mitigation Approach

Mitigation Approach	Mitigate	Defer	Accept
Pool	Pool 1	Pool 2	Pool 3

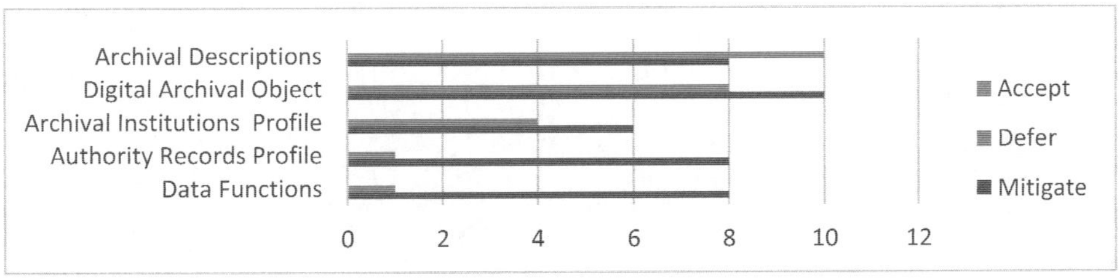

Figure 2. Mitigation Approach Results.

4. Conclusion

The results of risk assessment can support material to make risk management policies in preparing strategic planning of important electronic archive information security. It also can be used in developing comprehensive mitigation measures appropriate to the security of information assets that are archive from threats and disturbances that have occurred, are, or will occur. By implementing a risk assessment, SIKN JIKN can maintain and improve information security by knowing the potential threats, disturbances, and vulnerabilities to reduce losses on archives' information assets. Identifying risk assessments from 64 areas of concern resulted in an output in which 40 areas of concern had to mitigate, while 24 areas deferred.

Aknowledgements

This research publication was funded by the University of Indonesia's PUTI grant program (contract number: NKB-1079/UN2.RST/HKP.05.00/2020).

References

[1] Meng T and Hui L 2017 Application of Information Technology in Digital Archives Management *Proceedings of the 2017 International Conference on Education and E-Learning - ICEEL 2017* the 2017 International Conference (Bangkok, Thailand: ACM Press) pp 81–3
[2] Act No. 43 of 2009 about Archives (Indonesia)

[3] Quintana, A. and Nazar, M., 2020. [online] Ica.org. Available at: <https://www.ica.org/sites/default/files/the_role_of_archives_in_the_covid_19_crisis.pdf> [Accessed 6 July 2020]
[4] Grand Design of SIKN JIKN 2020-2024 (Indonesia)
[5] Bearman D Moments of Risk: Identifying Threats to Electronic Records 32
[6] Caralli R A, Stevens J F, Young L R and Wilson W R 2007 *Introducing OCTAVE Allegro: Improving the Information Security Risk Assessment Process:* (Fort Belvoir, VA: Defense Technical Information Center)
[7] Mohamad Ramli A A and Kadir Wan Dollah W A 2010 Threats to electronic records in Malaysian government organizations *2010 International Conference on Science and Social Research (CSSR 2010)* 2010 International Conference on Science and Social Research (CSSR) (Kuala Lumpur, Malaysia: IEEE) pp 228–33
[8] R. Assessment, "Risk Management: Implementation principles and Inventories for Risk Management/Risk Assessment methods and tools 'Survey of existing Risk Management and Risk Assessment Methods') Conducted by the Technical Department of ENISA Section Risk Management," no. June, 2006
[9] Jouini M, Rabai L B A and Aissa A B 2014 Classification of Security Threats in Information Systems *Procedia Computer Science* **32** 489–96

Forecasting of Freight Demand Transportation (Case Study in Lampung Province, Indonesia)

T Junaedi[1,2]

1 Civil Engineering and Environment Department, Institut Teknologi Bandung, Jl. Ganesha no.10 Bandung, Indonesia
2 Civil Engineering Department, Universitas Lampung, Jl. Sumantri Brojonegoro No.1 Gedong Meneng, Bandar Lampung, Indonesia, 35145

Email: tasan.junaedi@eng.unila.ac.id

Abstract. Forecasting the volume of freight has a significant importance for future decisions for the entire economy, as well as its individual participants. The transportation of goods between Sumatra and Java using the highway mode that accumulate with local transportation will greatly burden roads in Lampung province. This research was conducted to predict the pattern and amount of movement of goods in the future. The modelling is conducted for the plan period up to 20 years based on growth predictions of geographical, social, economic and traffic flow. This study conducted a process of Sequential Demand Model (Planning Model 4 stages) by using software Transplan Version 8. Transportation patterns in 2040 is predicted will be dominated by continuous trips with domination of movement from Java to South Sumatra, or otherwise. Both the transportation between internal zones and external zone to internal zone have a significant increase. The most strategic main line freight in Lampung for today and in the future is along the Bakauheni-South Sumatra and Bengkulu-South Sumatra both through toll roads, national roads, and railways.

1. Introduction

Forecasting the volume of freight has a significant importance for future decisions for the entire economy, as well as its individual participants. The size of transport is dependent on various factors and therefore more accurate forecasts require complex forecasting tools. However, for initial assessment the use of older simple models is sufficient; for example, Holt-Winters double exponential smoothing. They are widely used today in many areas of science, because of ease ofuse and low values of forecast errors. However, to this day such methods are still undergoing refinement. In article [1] the authors justify the need for adjustments in the Holt-Winters double exponential smoothing. In article [2] the authors present a selection method of parameters of model exponential smoothing, which are solved by the minimization of the problem. In [3] the authors of the most commonly used techniques for smoothing such as moving average or exponential smoothing. They specially designed weight coefficients which were employed for smoothing the forecast.

Currently, forecasting models are often supported by artificial intelligence methods. In article [4] the authors applied immune algorithm for estimating the optimal coefficients of logarithm support vector regression. In [5] neural networks of the Autoregressive Integrated Moving Average (ARIMA) method was used to reduce the sensitivity to input errors. In [6] the Bayesian network was used to predict the stock price.

The future of road transportation development impacts investment decisions of companies. An interesting approach to the use of forecasting in the supply chain optimization is described in [7]. The article presents the optimization of the supply chain cost by methods of integer programming. Data needed to optimize the production capacity and warehouse inventory is obtained by forecasting using the method of exponential smoothing. A lot can be found in literature on the use of forecasting in the fields of transport. A guidebook [8] has been compiled for those involved in transport planning. It provides a number of forecasting techniques.

The purpose of this study was to determine the pattern and the amount of movement of goods in the base year and the next few years, therefore the needs of the facilities which must be provided can be known to anticipate these developments. This study used a period of up to 20 years, which means that the results of this study can be used up to 2040 with the stages are: 2020-2025, 2026-2030, 2031-2035 and 2036-2040.

2. Research Methodology

Conventional transport models was used for prediction of traffic demand. The modeling is conducted for a period of up to 20 years where the traffic load adjusted for origin-destination matrix of the associated zones that have been adapted to the level traffic growth of modeling results. The results of the modeling include the prediction of traffic volume and traffic performance predictions in the future.

A process Sequential Demand Model (Planning Model 4 stages) was done in this research. The use of this model is based on some considerations: firstly, it is easy to control on each stage of the work; secondly, it has higher accuracy related to the simplification process is very limited when compared with the Direct Demand Model; thirdly, this model is very flexible to be applied to new cities studied ; and finally, it is in accordance with the ability of a computer program provided by the TRANPLAN Version 8.

Steps to run the program for TRANPLAN as follows:

a) Creating and loading freight road network is not the route (Highway Network);
b) Make a trip freight distribution as input on the process of loading;
c) Imposition into the road network (Equilibrium highway Load);
d) Make transyt network, the network of public transport routes in the route);
e) Creating route network path (Build Path Transyt);
f) Creating origin-destination matrix with public transportation;
g) Imposition (Transyt Loading).

3. Data and Instrumentation

Several parameters were taken and adapted to the statistical report of the Central Statistics Agency (BPS) and the reports from the relevant authorities. Meanwhile, data from the primary survey more widely used in the process of validation and adjustment of multiple parameters.

The data used as input in the development of the model, are:

a) The data of land use and zoning administration as a basis for establishing the zone trip system
b) The data of Prior OD Matrix. In this case, using the data from the origin-destination survey, OD data in 2018, and secondary data from the results of previous studies;
c) The data of condition of transport infrastructure, including: rail networks, stations, road networks, traffic management system, as a basis for the development of the transport network models;
d) The data of population distribution, income, vehicle ownership, land use, PDRB, and data traffic as input for the attraction estimation and trip distribution as well as the validation OD matrix.

The study area is divided into several zones. For convenience, the zoning is based on administrative boundaries. Thus in this analysis defined 30 zones consisting of 27 internal zones and 3 external zone as shown in Figure 1.

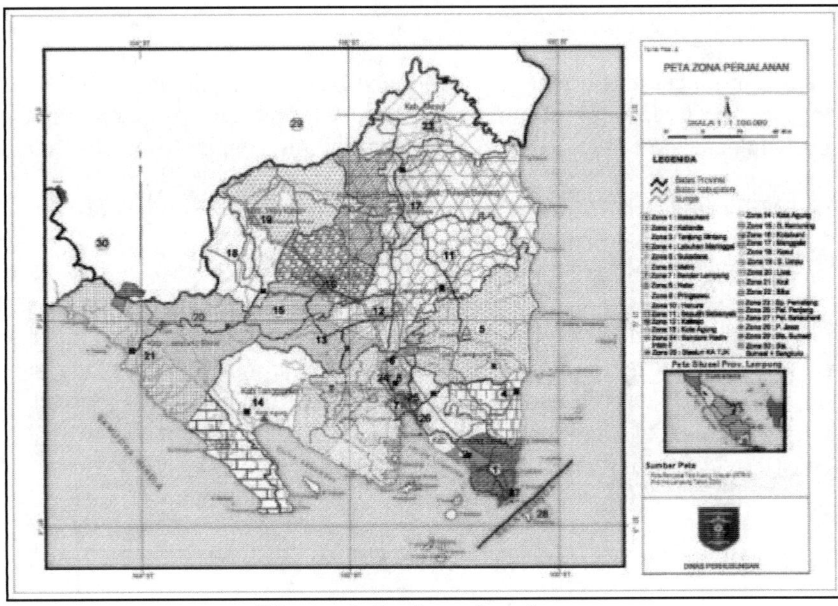

Figure 1. Zoning of study area

Lampung province road network is divided into 30 zones, 1758 nodes and 4580 roads. Roads were included in this study are arterial roads and collector roads. Inlet and outlet is divided into three, namely the entrance of Bengkulu, South Sumatra, and from Java. Lampung province road network model is in autocad format then was entered into the base data as input Transplant program (Figure 2).

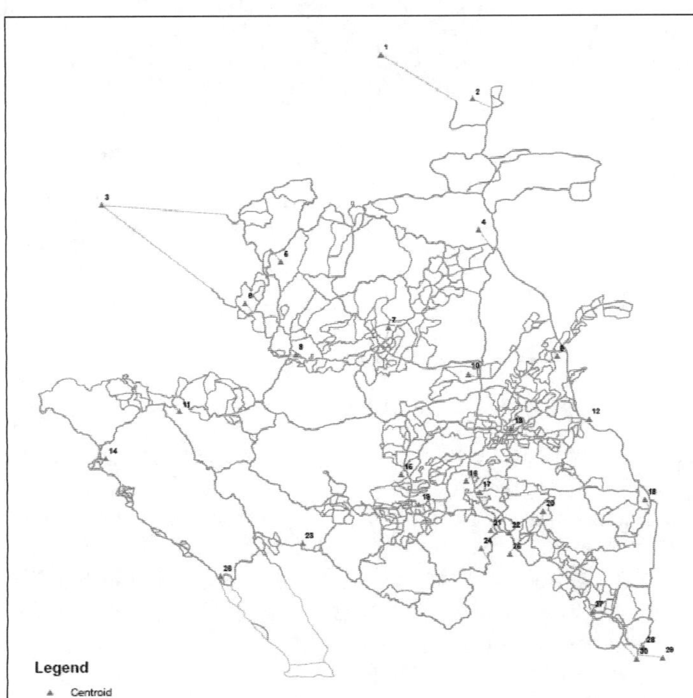

Figure 2. Road network in Lampung province

4. Results and Discussion

4.1. Freight Transportation in Base Year

In 2020 the average production of travel goods for each zone approximately 792 344 tons / day, with Bandar Lampung as the largest origin and destination of goods in Lampung province for 131 020 tons/day. In percentages, the movement in Bandar Lampung approximately 16.5% of the whole movement in Lampung province. Followed by Java amounted to 90,322 tons/day or 11.4%, and South Sumatra for 58 507 tons/day or 7.4% of the total freight movement in the province of Lampung and then Metro for 50 856 tons/day or approximately 6.4%.

Based on data generation and attraction, Origin Destination Matrix for goods is predicted for the year 2020 to 2040. This matrix noticed the vehicle travel patterns between regions. This matrix is processed from the origin destination survey (road side interview) and the results show that Bandar Lampung is the center of attraction which is become the destination from many surrounding areas. After that MAT is loaded to the road network in order to see the pattern of movement of passengers and goods in the road network.

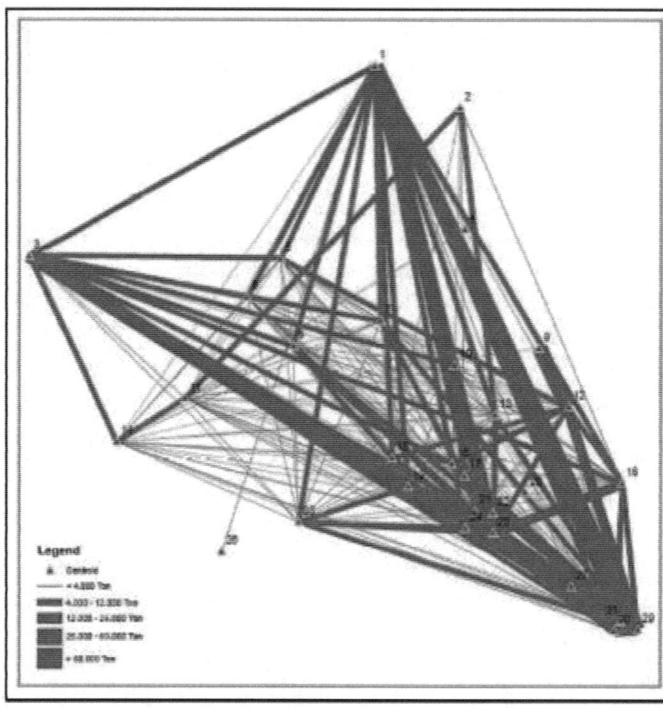

Figure 3. Desire line of transportation in 2020

Based on the analysis, show that the provincial capital of Lampung, Bandar Lampung (Zone 7) has a origin and destination of the trip dominantly, ie 131 020 and 77 148 tons/day. The transport in Lampung are also quite dominant come from Zone 28 (from Java through the Bakauheni Port) of 90322 tons/day and 65680 tons/day), South Sumatra for 58507 and 53599 tons/day, Metro with 50856 and 38976 tons/day, and Bakauheni with 42,118 tons/day and 45,564 tons/day.

Based on the analysis of the survey results of OD national and other sources, there is information on the distribution of travel demand of goods to every available moda, shown on the image above. The prediction of this the moda use is intended only for the freight movement overland via the internal network in the area in Lampung where external movement generally to Java and Sumatra uses road modas, while to all other areas by sea. It appears that the current freight transport is still dominated by trucks on the highway, while the railway transport only accounts for no more than 10%.

Ideally, with the railway network development scenarios, as well as the utilization of the rail line, in the future the freight transport needs should be more accommodated by train. This is because train is much more efficient than road transport with trucks in case of the time and expense for regional freight movement.

Reducing the burden of road network in the future for Lampung province is an important strategy. This is because the condition of existing the road network in the province within the next 10 years will face big problems such as traffic congestions. Therefore the railways development is a good solution to overcome this problem and also to encourage the development marine transport system scenarios.

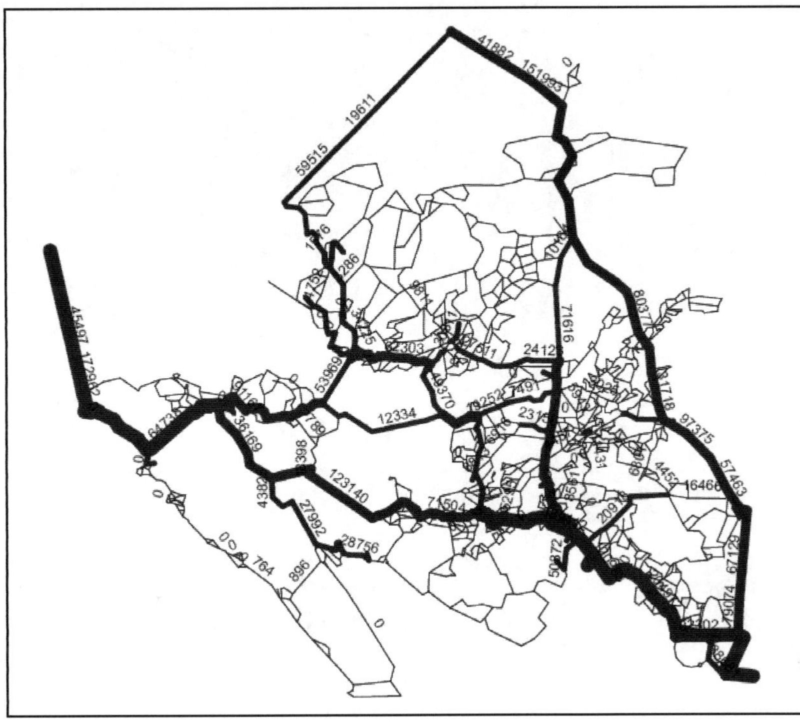

Figure 4. Results of loading to network

From the picture above can be seen that the most of goods movement by road from Bakauheni to Bandar Lampung then continue to Bengkulu, which amounted to 172 962 tons/day. While the movement of goods through Bakauheni to Bandar Lampung then continued until Menggala (Lintas Tengah) is approximately 139,412 tons/day. Movement of goods from Bakauheni through Lintas Timur is 97,375 tons/day. While the goods movement through road between Menggala to South Sumatra is 151,993 tons/day.

The imposition of provincial road network is relatively evenly and provide a relatively equal access to all parts of Lampung. That pattern can provide support for the achievement of room development which is more proportional, particularly for areas of the West Lampung and the East Lampung.

To assess the performance of the road network, use indicator of the volume of traffic per road capacity, which is known as Volume-Capacity Ratio (VCR).

Figure 5 shows that currently some segments of Lintas Timur, from South Sumatra-Menggala-Tulang Bawang, have LOS C. While, the segments of Lintas Timur (Sukadana-Bakauheni), Lintas Barat (Kota Agung-Bandar Lampung), dan Lintas Tengah (Bandar Jaya-Bandar Lampung-Kalianda), have LOS B.

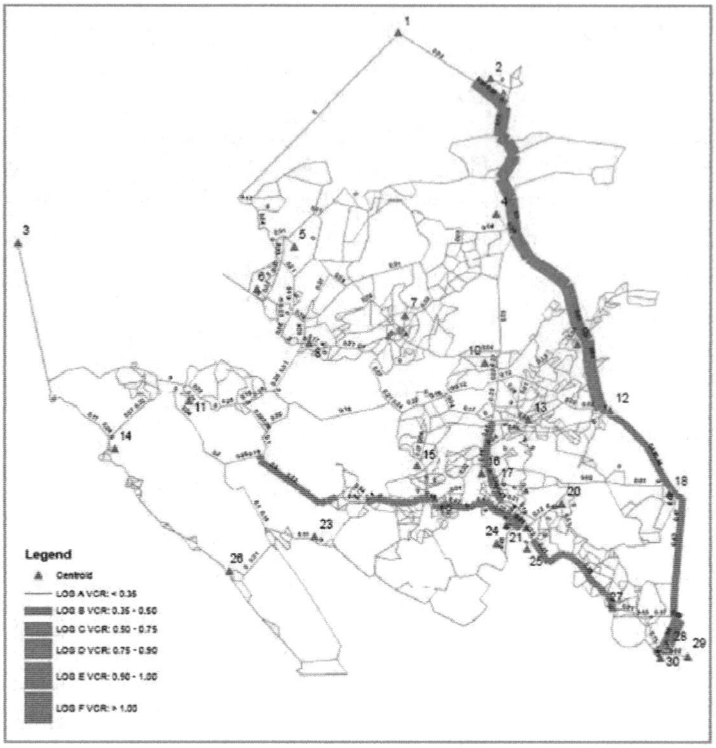

Figure 5. Volume Capacity Ratio in 2020

4.2. Prediction of Transportation

The movement of goods is projected for the next 5 years and so on up to the time of planning, which is 30 years old. Based on this projection could be seen the amount of movement which is happened, so that if it is charged to the road network, the road network that need to be addressed will be known, as well as its relation to other modas. From the desire line in 2025 can be seen the amount of movement that occurs increased about 1.2 times of the movement in 2020. But the movement pattern was relatively similar with patterns in 2020.

The amount of continuous movement is going greater with the domination of the movement of Java-Sumatera, or otherwise. The amount of movement between the zones of internal and external zone to the internal zone is also seen to increase. This will burden the existing road network in the province of Lampung.

Existing trip patterns are also predicted for the next few years, and it can be seen that the increase of movement amount happen on every street. The pattern has not changed much, but the increase of amount of movement can be seen.

In 2025 the movement pattern is predicted to increase with biggest road loading on link of Bakauheni - Bandar Lampung , amounting to 393.606 tonnes/day, followed by Tanggamus towards Liwa for 393.274 tonnes/day. The movement of goods through Lintas Timur is at 265.665 tonnes / day, up to the road to South Sumatra. Another problem faced is the movement of regional which across Lampung province. The high volume of regional movements with the origin and destination regions outside Lampung has created road damage in Lampung province.

In 2030, the movement patterns of freight is predicted to increase with biggest road loading on link of Bakauheni - Bandar Lampung at 496 402 tonnes/day, which has an increase of 100.000 tons / day from the previous 5 years. Followed by a link of Tanggamus to Liwa at 383.463 tonnes/day. The movement of goods through Lintas Timur is accounted for 308.861 tonnes/day, up to the road to South Sumatra.

IOP Conf. Series: Materials Science and Engineering 1232 (2022) 012008 doi:10.1088/1757-899X/1232/1/012008

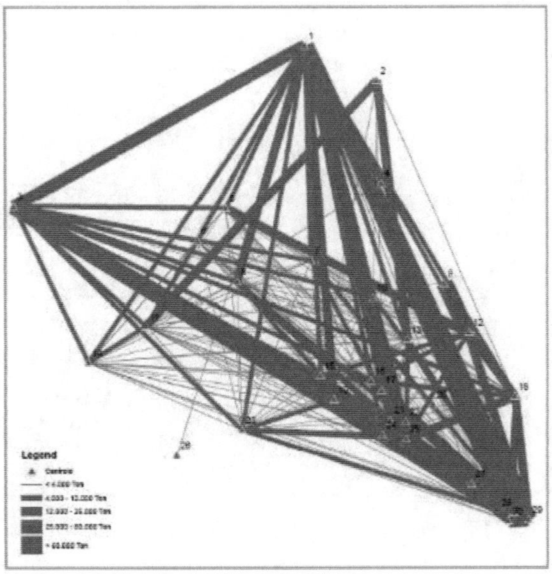

Figure 6. Desire line in 2025

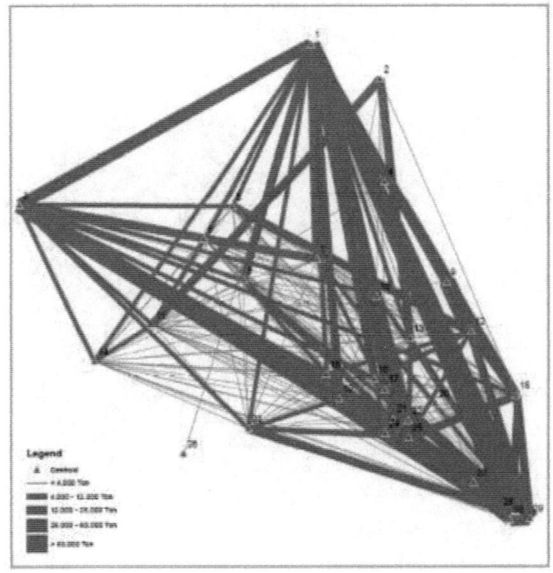

Figure 7. Desire line in 2030

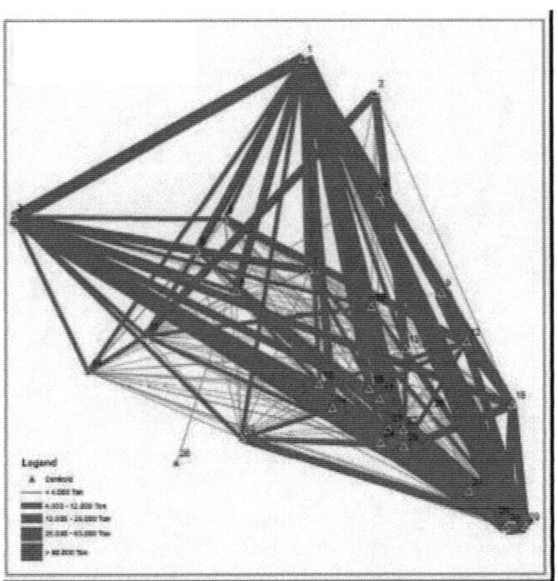

Figure 8. Desire line in 2035

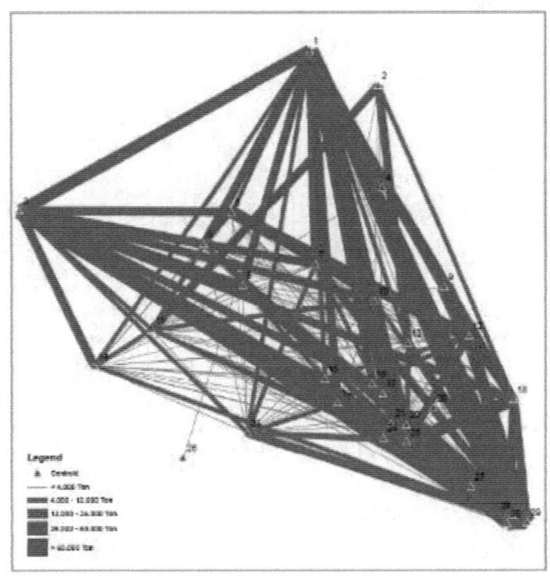

Figure 9. Desire line in 2040

In 2035, the biggest movement of goods is predicted similar with the prediction of the previous year, namely on the link Bakauheni - Bandar Lampung ranged from 425.193 – 564.039 ton/day. The movement of goods from Bandar Lampung towards Bengkulu ranged from 399.134 - 435.976 tonnes/day. While the movements through Lintas Timur to Sukadana is 224.863-353.709 tonnes/day, and from Sukadana to South Sumatra is accounted for 251.370-384.772 tonnes/day.

At year end 2040, the pattern of movements is still have a same trend that the biggest movement through the link of Bakauheni - Bandar Lampung ranged from 483,418 to 641,280 tonnes/day. From Bandar Lampung towards Bengkulu ranged from 428,528-473,544 tonnes/day. While the movements through Lintas Timur until Sukadana is accounted for 322,436-402,145 tonnes/day, and from Sukadana up the road to the South Sumatra ranged from 333,346 to 437,462 tonnes/day.

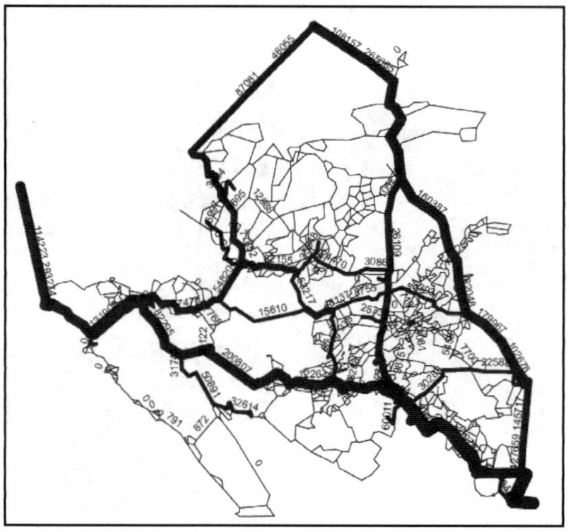

Figure 10. Loading to network in 2025

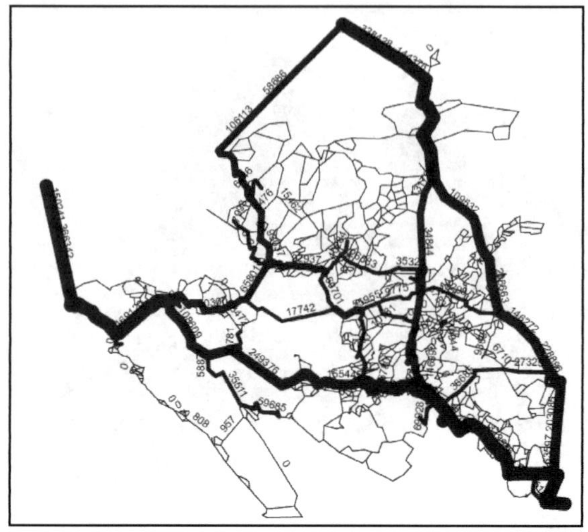

Figure 11. Loading to network in 2030

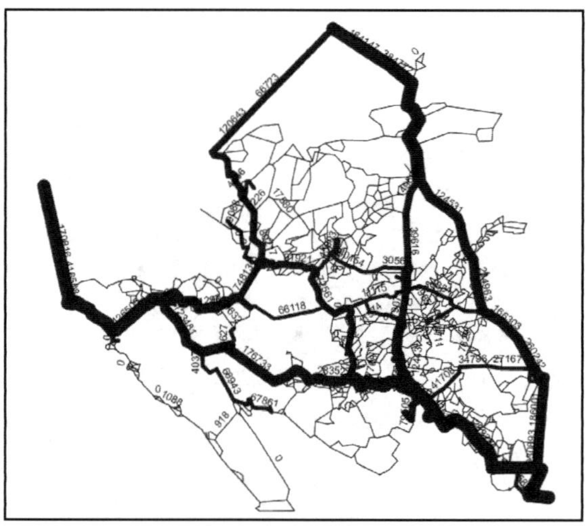

Figure 12. Loading to network in 2030

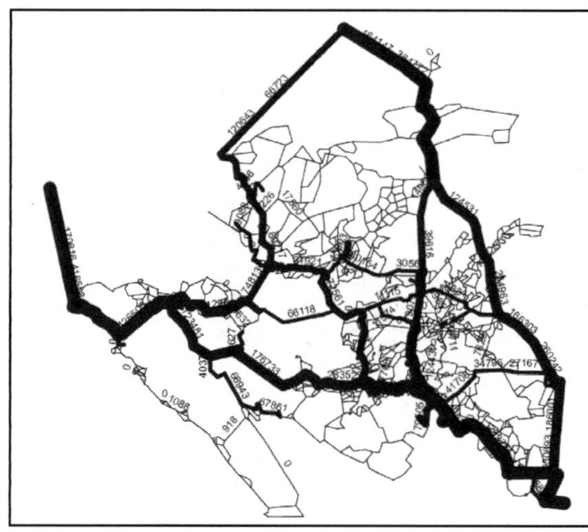

Figure 13. Loading to network in 2035

Several roads are in good condition so it can be passed by heavy vehicles without causing any damage. While some other roads were only able to be passed by vehicles which have axle loads is not too heavy, as a result, the transport of goods by heavy vehicles can only be served by certain road sections. In several sections, the traffic network is passing through the unfavorable geographical conditions and it needs alternative pathways to accelerate the trip.

Based on the characteristics of the transport of goods, it consists of industrial and agricultural / plantation transportations which use the moda of medium truck and pick-up. This indicates that the the transport of goods use vehicles which have medium axle load. This is because the center of industrial and agriculture/plantations are the areas that can only be reached by small-medium vehicles.

The condition above gives an overview of types of freight transport required by Lampung province, namely the transport of goods by vehicles having small-medium capacity and significant numbers (as evidenced by the amount of movement frequency is every day), so that it can reach out to all parts of Lampung. This provides an advantage, because network traffic required must not use the road with high bearing capacity, but the road can serve small-to-medium vehicles well and can reach out to all corners of the region.

IOP Conf. Series: Materials Science and Engineering 1232 (2022) 012008 doi:10.1088/1757-899X/1232/1/012008

4.3. Road Network Performance

The performance of the road network in the prediction of the year 2025 will be clearly shown in the figure. VCR value of each road segment as presented in Figures 14 to 17.

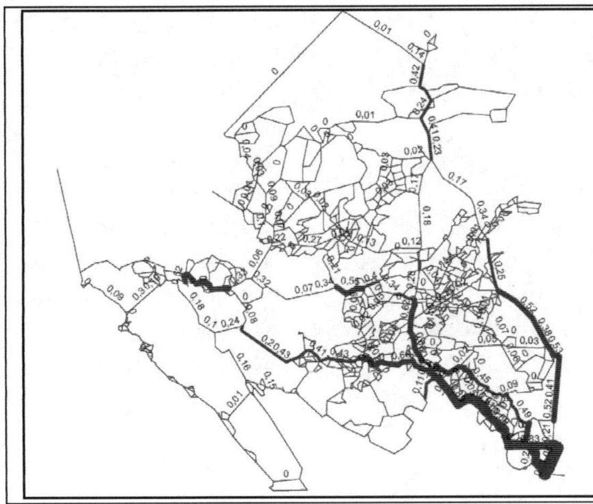

Figure 14. Value of VCR in 2025

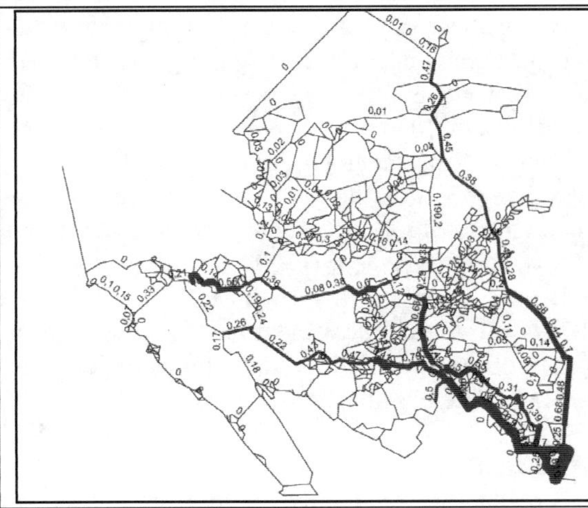

Figure 15. Value of VCR in 2030

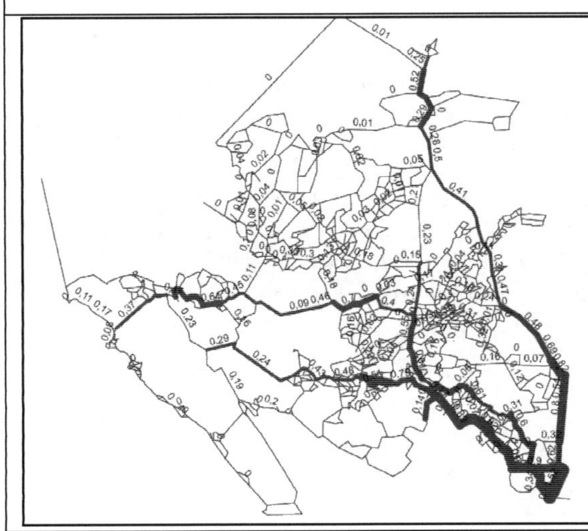

Figure 16. Value of VCR in 2030

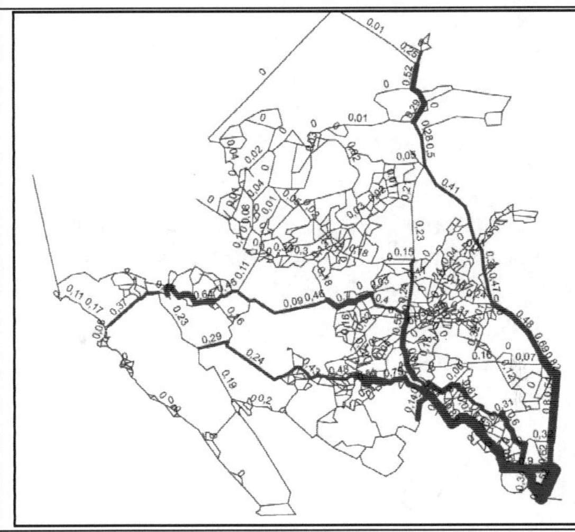

Figure 17. Value of VCR in 2035

It can be seen from the figure that the VCR value for the segment of Bakauheni to Bandar Lampung through the existing road is 0.89. Meanwhile, if the toll road is built, then the value of the VCR is turned into around 0.49. Value VCR roads connecting Bandar Lampung towards Pringsewu also increase to 0.66, as well as through Lintas Timur for 0.53. We can also see that the movement of traffic from Terbanggi towards Kotabum has VCR value by 0.55 and from Sumberjaya towards Liwa by 0.52. In addition, the value of VCR from Menggala towards Mesuji around 0.42.

In 2030 the movement of goods is predicted to spread evenly with the value of VCR of Bakauheni towards Bandar Lampung ranged from 0.81 to 0.9. VCR value for the segment towards Bandar Lampung Tanggamus around 0.79. The value of VCR for the transport in Lintas Timur varied by 0.44 to 0.7. The substantial movement is also evident in link of Bandar Jaya towards Padang Ratu with VCR value of about 0.6, as well as the road linking Kenali and Liwa has the traffic volume at 0.56

The increase of movement in 2035 can be seen more clearly with the indicator value of VCR (road performance) as presented in Figure 17.

The biggest changes of the value of VCR between 2030 and 2035 is the freight movement of Bakauheni towards Bandar Lampung which ranged from 0.9 to 1.52. This shows the road has been in a critical condition, eventhough some of traffics is divided trough toll road with VCR value of 0.61. In 2035, the movement of goods through the Lintas TImur has VCR value varying from 0.62 to 0.80.

While the roads of Bandar Lampung towards Tanggamus still has VCR value around 0.79. A substantial movement is also evident from Bandar Jaya towards Padang Ratu with VCR value of about 0.7. As well as road connecting Kenali and Liwa has traffic volume around 0.64, or has a slight increase from 2030.

The biggest changes of the value of VCR between 2035 and 2030 is the freight movement of Bakauheni towards Bandar Lampung which ranged from 1.0 to 2.27. This shows the road has been in a stagnan condition, eventhough some of traffics is divided trough toll road with VCR value of 0.80. In 2040, the movement of goods through the Lintas Timur has VCR value varying, with the highest VCR is 0.87. While the roads of Bandar Lampung towards Tanggamus still has VCR value around 0.82. A substantial movement is also evident from Bandar Jaya towards Padang Ratu with VCR value of about 0.91. As well as road connecting Kenali and Liwa has traffic volume around 0.89, or has a significant increase from 2025.

5. Conclusions

The conclusions that can be drawn from this study are as follows:

1) Most of the movement of goods in the base year (2020) through the national road from Bakauheni to Bandar Lampung, and continue to the Central Lampung were then divided into two directions, namely through Lintas Tengah and Lintas Timur toward South Sumatra.
2) The movement of freight in Lampung Province (both between internal zones and external zones) is dominated by moda of trucks on the highway, while the railway transport only accounts for no more than 10%.
3) The pattern of freight movement in 2040 is predicted that the domination of continuously movements from Java to Sumatera or otherwise will be greater. The number of movements both between internal zones and between internal and external zones will increase significantly.
4) Until 2040, the transport of goods from Java to Sumatra (inter-urban) were still dominant. It can be concluded that the main line of freight transport in Lampung, which is most strategic for now and in the future, is along the Bakauheni to South Sumatra and also Bengkulu either through toll roads, national roads, and railways.

.

6. References

[1] R. Lawton, “How should additive Holt–Winters estimates be corrected?” *Int, J. Forecast,* 14(3):393-403, 1998. doi:10.1016/S0169-2070(98)00040-5

[2] R. Pinto, P. Gaiardelli, “Setting forecasting model parameters using unconstrained direct search methods: An empirical evaluation,” *Expert Syst. Appl*, 40(13):5331-5340, 2013. doi:10.1016/j. eswa.2013.03.044

[3] RR. Yager, “Exponential smoothing with credibility weighted observations. *Information Sciences*,” 252:96-105, 2013. doi:10.1016/j.ins.2013.07.008

[4] K-P Lin, P-F. Pai, S-L. Yang, “Forecasting concentrations of air pollutants by logarithm support vector regression with immune algorithms,” *Appl. Math. Comput,* 217(12):5318-5327, 2011. doi:10.1016/j. amc.2010.11.055

[5] B. Lénárt, “Automatic identification of ARIMA models with neural network,” *Period. Polytech, Transp*, 39(1):39-42, 2011. doi:10.3311/pp.tr.2011-1.07

[6] K. Eisuke, H. Maasaki, M.Takao, “Application of Bayesian Network to stock price prediction,” *Artif. Intell. Research*, 1(2):171-184, 2012.

[7] Y. Acar, E.S. Gardner, “Forecasting method selection in a global supply chain,” *Int. J. Forecast*, 28(4):842– 848, 2012. doi:10.1016/j.ijforecast.2011.11.003

[8] A guidebook for forecasting freight transportation demand. Cambridge Systematics, Inc. 1997; NCHRPreport 388, Transportation Research Board, National Research Council, Washington, D.C.

IOP Conf. Series: Materials Science and Engineering 1232 (2022) 012009 doi:10.1088/1757-899X/1232/1/012009

Enhancing the flexural load capacity of the reinforced concrete beams using Dramix 3D 80/60 BG steel fibres

V A Noorhidana[1], M Isneini[2], A M Putra[3]

Civil Engineering Department, Universitas Lampung,
Jl. Sumantri Brojonegoro No.1 Gedong Meneng, Bandar Lampung 35145, Indonesia

Email: [1] vera.agustriana@eng.unila.ac.id, [2] isneini.m@gmail.com, [3] asyrofimiranda97@gmail.com

Abstract. The purpose of this study was to investigate the effect of adding Dramix 3D 80/60 BG steel fibres to the reinforced concrete beams experimentally. The main specimens were 6 reinforced concrete beams (dimension of 150 x 180 x 1700 mm) with a variety of the flexural reinforcement (i.e. 2D6, 2D10 and 3D10) and the fibre volume fraction (i.e. 0%, 0.5%, 1% and 1.5%). The concrete compressive strength was 25 MPa, no superplasticizer was added in the concrete mix. The test set-up of the beam specimen is, the beam was simply supports at the distance of 1500mm, 2 concentrated loads gradually were applied until it collapses. During applying the loads, measurements were taken including the deflection in the mid-span of the beam, the crack propagation and crack width. The results showed that the increase of fibre content will increase the load capacity and stiffness without reducing the ductility of the reinforced concrete beams. The steel fibre volume fraction of 1% is the maximum content that can be used without using superplasticizer to maintain the workability of the fresh concrete. The addition of 0.5% steel fibre provides a similar flexural performance to the addition of one D10 (flexural reinforcement) in a reinforced concrete beam.

1. Introduction

Reinforced concrete structures consist of vertical structural elements (or called columns) and horizontal structural elements (also called beams). Beams experience bending due to loads acting on them. In the reinforced concrete beam collapse theory, there are 2 types of collapse, namely ductile collapse and brittle collapse. Reinforced concrete beams must be designed to experience ductile collapse by limiting the flexural reinforcement ratio (ρ) between ρ_{min} and ρ_{max}, or it is called under-reinforced. Ductile collapse in a beam is a typical safe collapse because the beam will show physical signs such as substantial deflection and cracks before collapse. Conversely, if the flexural reinforcement ratio is over-reinforced ($\rho_{used} > \rho_{max}$) then the beam will experience a brittle collapse which is dangerous because it is a sudden collapse without showing any physical signs.

Research on the use of fibre reinforced concrete (FRC) in reinforced concrete beam structures has been widely carried out. Some researchers [1,2,3,4] have studied the effect of using steel FRC (SFRC) on the flexural behaviour of reinforced concrete beams. Several researchers have also made an approach model to analyze the strength of reinforced concrete beams using the finite element method [5, 6]. The results of these studies indicate that SFRC can improve the flexural performance of reinforced concrete beams using the flexural reinforcement ratio (ρ) is between ρ_{min} and ρ_{max}, or called under-reinforced.

From the results of previous studies, it is not known how the effect of SFRC on the behavior of reinforced concrete beams with flexural reinforcement less than ρ_{min} or greater than ρ_{max}. This information is important to know, for example in the reinforced concrete beams repair with the method of enlarging the dimensions of the beam section which will also increases the amount of used flexural reinforcement. The addition of this reinforcement also increases the flexural reinforcement ratio. If not careful, then the flexural reinforcement ratio after repair is greater than ρ_{max} which will resulting a brittle collapse.

Therefore, it is necessary to anticipate so that even though there is excessive flexural reinforcement, it still performs a ductile flexural behaviour. One of them is by combining it with steel fibre reinforced concrete (SFRC). Steel fibre reinforced concrete is a mixture of ordinary concrete which steel fibres was added in the mixing process of the concrete, so that when the concrete hardens, the steel fibres are scattered randomly and evenly in the concrete, which can withstand the tensile stress in the concrete in any directions. SFRC has advantages over ordinary concrete, including having higher tensile strength, ductility, resistance to impact loads, and a modulus of elasticity in comparison with ordinary concrete, and has a lower shrinkage rate than ordinary concrete [7].

The main objective of this study is to investigate the flexural behaviour of SFRC beams using various volume fraction of Dramix 3D 80/60 BG steel fibres and the flexural reinforcement of the beam, in terms of the load-deflection relationship, ductility and beam failure pattern experimentally.

2. Experimental Program

The main specimens are reinforced concrete beams with the dimensions of 150 x 190 x 1700 mm. Detailed dimensions and reinforcement of the test specimens are presented in Figure 1 and the details of the test specimens variation are shown in Table 1.

The beam specimens were cast with the concrete strength of 25 MPa. The yield strength of steel reinforcement was 380 MPa. Dramix 3D 80/60 BG steel fibre was used in this study with the variations of 0%, 0.5%, 1%, 1.5%. The variation of the flexural reinforcement of the beam specimens were 2-D6, 2-D10 and 3-D10. The reinforcement of 2-D10 is identified under the minimum reinforcement requirement. The beams with 2-D10 and 3-D10 were lied within the required reinforcement range. These beams are designed to experience flexural failure. The material composition for plain concrete and concrete with various steel fibre volume fraction is shown in Table 2.

The beam specimen tests were carried out at 28 days of concrete. The beam specimen was placed on 2 supports at a distance of 1500mm on a loading frame, then 2 concentrated loads vertically were applied at the distance as shown in Figure 2, until the beam collapses. In the middle of the beam span, a dial gauge was installed to measure the deflection of the beam. During the loading process, the measurements including the deflection, crack development and crack width according to the load increase were recorded.

To measure the mechanical properties of concrete, cube samples (150 x 150 x 150) and prism (100 x 100 x 600 mm) were also made to determine the compressive strength and flexural strength of concrete (modulus of rupture). The tests were conducted according to ASTM C39 and ASTM C496. They were tested at the same day with the testing of the beam specimens.

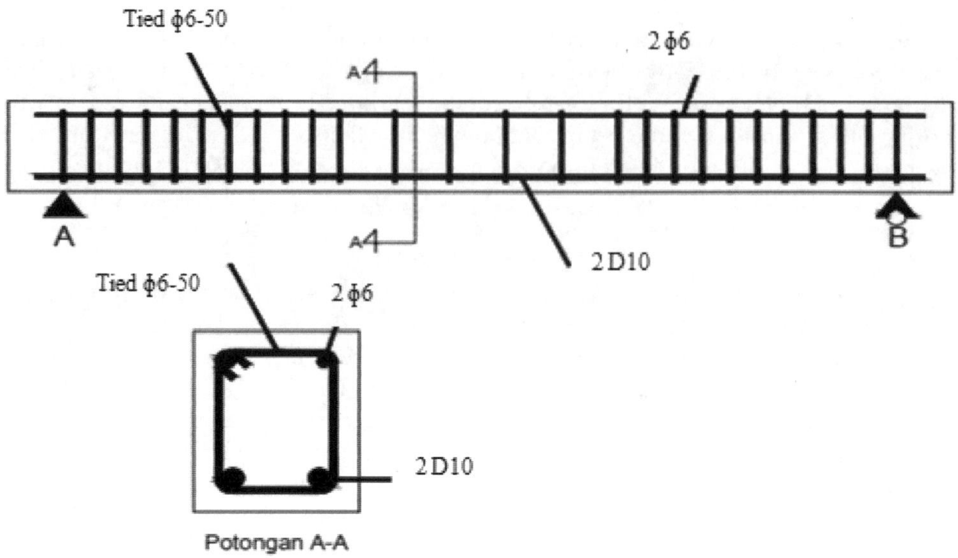

Figure 1. The beam specimen detail

Figure 2. The test set-up of the beam specimen

IOP Conf. Series: Materials Science and Engineering 1232 (2022) 012009 doi:10.1088/1757-899X/1232/1/012009

Table 1. Detail reinforcement of the reinforced concrete beam.

Beam specimen	Flexural reinforcement	ρ	ρ_{min}	Tied	Steel fibre volume fraction, V_f (%)
BL1	2-D10	0.0066	0.0037	ϕ6-50	0
BL2	2-D10	0.0066	0.0037	ϕ6-50	0.5
BL3	2-D10	0.0066	0.0037	ϕ6-50	1
BL4	2-D10	0.0066	0.0037	ϕ6-50	1.5
BL5	3-D10	0.0099	0.0037	ϕ6-50	1
BL6	2-D6	0.0023	0.0037	ϕ6-50	1

Table 2. Material for plain/steel fibre reinforced concrete (per m^3)

Steel fibre volume fraction, V_f (%)	Cement (kg)	Sand (kg)	Coarse aggregate (kg)	Water (kg)	Steel fibre (kg)
0	375	768	902	225	-
0.5	363,771	745,003	874,990	218,262	39,25
1	362,005	741,386	870,743	217,203	78,50
1.5	360,256	737,805	866,536	216,153	117,75

3. Test results

3.1. Mechanical properties

Table 3 and Figure 3 and 4 provide data on the effect of adding steel fibres to the concrete workability (slump test), compressive strength and flexural tensile strength (modulus of rupture) of the concrete.

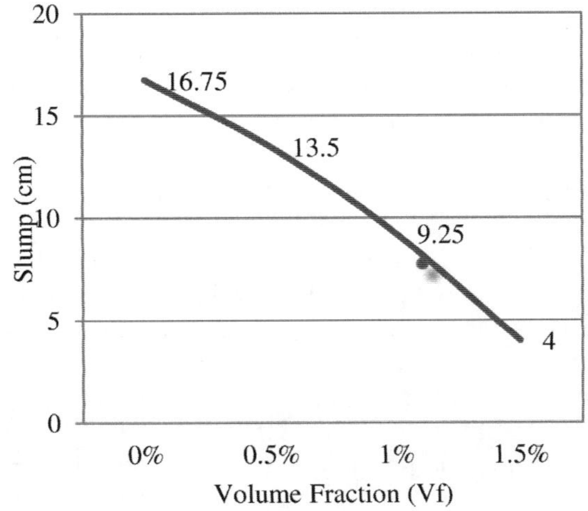

Figure 3. The slump - fibre volume fraction relationship

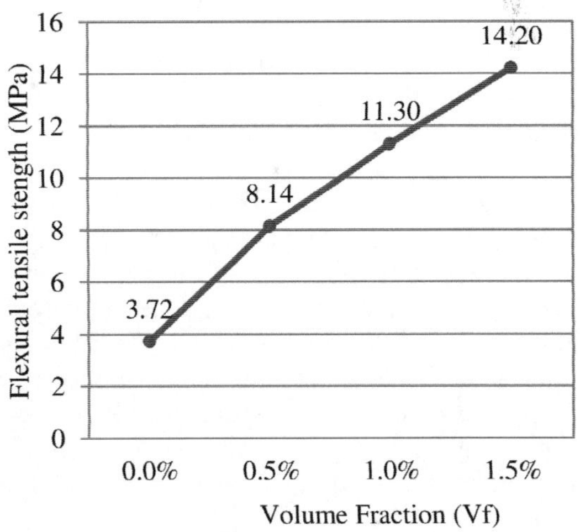

Figure 4. The flexural strength –fibre volume fraction relationship

Table 3. Slump, average of compressive strength and flexural tensile strength

Steel fibre volume fraction, Vf (%)	Slump (cm)	Ave. compressive strength			Ave. flexural tensile strength	
		Cube strength (N/mm^2)	Covert to cylinder strength (N/mm^2)	Increase (%)	Flexural tensile strength (N/mm^2)	Increase (%)
0	16,75	30,54	25,35	0	3,72	0
0.5	13.50	28,58	23,72	-6,43	8,14	118,82
1	9.25	33,06	27,44	5,39	11,30	203,76
1.5	4.00	31,93	26,50	4,53	14,20	281,72

Workability of the concrete mix determines the ease of mixing, transporting, molding, compaction, and finishing. Generally concrete slump value (obtained from Slump Test) is used to find the workability. The higher the slump value, the easier the workability. The workability is determined by several factors, such as the water-cement ratio, the amount of water used, the aggregate condition (wet, dry or saturated surface dry). In this study, the design slump values for normal concrete ranged from 60-180 mm. The slump value of concrete/fibre reinforced concrete for each concrete mix is presented in Table 3 and Figure 3.

It can be seen in Table 3, the greater fibre volume fraction, the lower the concrete slump value. This is because the steel fibres in the concrete mix lock the coarse aggregate, thereby reducing the movement of the coarse aggregate. In addition, the water in the concrete mix, besides lubricating the surface of the aggregate, must also lubricate the surface of the steel fibres so that the free water content in the concrete mix is reduced.

Based on the above results, it can be seen that the addition of the steel fibres up to $V_f = 1\%$ results in a slump value that still meets the design slump of 60-180 mm. However, at $V_f = 1.5\%$, the slump value no longer meets the requirements, which means that the workability of the concrete mix is very low. This is not good because it will reduce the solidity of the concrete; the concrete has many porous.

To overcome this, the use of superplasticizer is unavoided when using fibre volume fraction $V_f \geq 1.5\%$, this is to improve the concrete workability so that the design slump can be achieved.

Based on the Table 3, it can be seen that the addition of steel fibres to the concrete mix does not have a significant effect on the increase in the concrete compressive strength, which is only around 0 - 10%.

The presence of steel fibres in the concrete has a significant effect on the failure mode during the compression test. The plain concrete has collapsed by releasing some parts of the sample cube. Meanwhile, the fibre reinforced concrete cube sample only suffered cracks without any loose concrete parts. This is because the steel fibres that are randomly scattered in the concrete hold/bind the coarse aggregates to each other so that they do not release when experiencing tensile stress. This shows that the addition of steel fibres to the concrete can change the concrete become more ductile material.

Figure 4 shows the concrete flexural strength increases with increasing the fibre volume fraction. The highest flexural strength of concrete is found in the volume fraction of 1.5%, which is 14.20 MPa (an increase of 281.42% from the plain concrete). This shows that the addition of steel fibre to the concrete mixture can increase the tensile strength of concrete.

The curve in Figure 4 continues to increase. It is not known whether after $V_f = 1.5\%$, the curve will continue to increase. This is possible, but the problem is when adding $V_f = 1.5\%$, of steel fibre, the concrete mix become stiffer and nor workable, so that the superplasticizer is needed to solve the workability problem.

In the plain concrete, the tensile stress that occurs is only held by the concrete itself, while in the fibre reinforced concrete the tensile stress that occurs is held together by the concrete and steel fibres. This is why the flexural strength of the steel fibre reinforced concrete is higher than those of the plain concrete.

3.2. Reinforced concrete beam performance

Table 4 and Figure 5-7 show the data from the reinforced concrete beam testing, which shows the effect of the steel fibres content on the maximum load capacity, the load-deflection curve of the beam, the crack-pattern and the crack width.

As a result of the applied loading, all beam specimen experienced flexural collapse. The first crack occurred between the 2 point loads and then propagated vertically. Then cracks formed again along the centre-span of the beam. The crack pattern of each beam specimen was recorded and the results are shown in Figure 5.

Table 4 shows the effect of steel fibre volume fraction (V_f) on the first crack load, the yield load, the ultimate load and the crack width. It can be seen that those loads increased with the increasing of steel fibre content mixed into the concrete. This means that steel fibres in the concrete improved the load capacity of the beam.

The presence of steel fibres that are randomly scattered in the concrete reduces the crack widths of the cracks that occur, as shown in the load - crack width relationship in Figure 6. The more amount of fibre incorporating the concrete, the smaller crack-width occurs. This is because the steel fibres bridge the cracks that occur thus slowing the rate of propagation of the cracks, which in turn increases the load capacity of the beam.

The crack patterns of the beam specimens with no-fibre and with fibre look identical, which is starting with the formation of flexural cracks and them flexural-shear cracks leading to the supports. However, the beam specimen with no-fibre experienced more cracks.

Table 4. Loads, deflection and crack width of the beam specimens

Specimen		Load (kg)	Increase (%)	Deflection (mm)	Crack width (div)
BL 1	First crack	1108.78		0.42	-
	Yield	3191.95		4.54	-
	Ultimate	3796.74	-	14.57	-
BL 2	First crack	1175.98	6.06	0.49	2
	Yield	3326.34	4.21	4.83	38
	Ultimate	4065.53	7.08	17.31	131
BL 3	First crack	1579.17	42.42	0.60	4
	Yield	4065.53	27.37	5.02	23
	Ultimate	4603.12	21.24	16.84	112
BL 4	First crack	2452.76	121.21	1.05	5
	Yield	5543.91	73.68	7.67	62
	Ultimate	5947.10	56.64	17.13	108
BL 5	First crack	1982.37	78.79	0.79	2
	Yield	5947.10	86.32	16.03	98
	Ultimate	6081.50	60.18	17.82	104
BL 6	First crack	1175.98	6.06	1.32	4
	Yield	2049.57	-35.79	4.22	58
	Ultimate	2318.36	-38.94	16.99	118

IOP Conf. Series: Materials Science and Engineering 1232 (2022) 012009 doi:10.1088/1757-899X/1232/1/012009

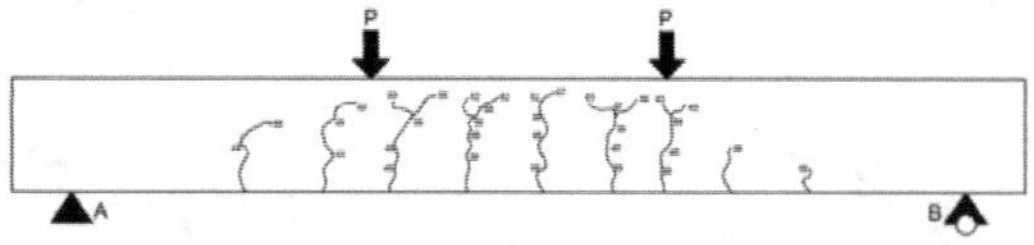

a. BL1 (2-D10, V_f= 0%)

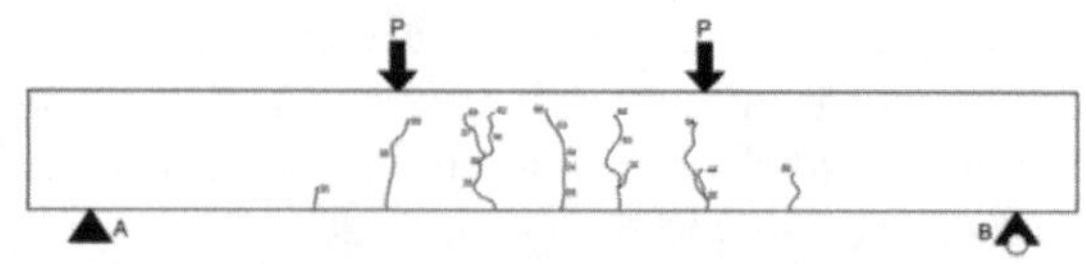

b. BL2 (2-D10, V_f= 0.5%)

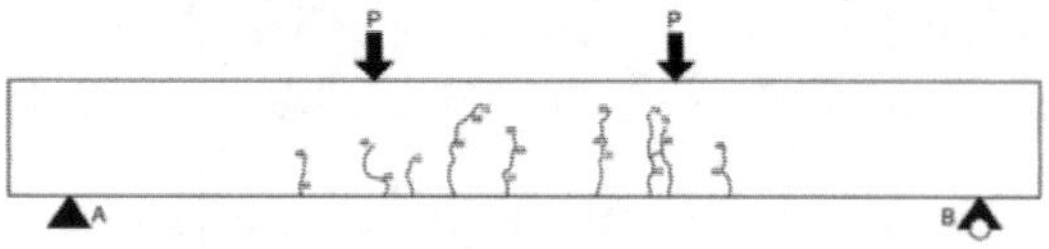

c. BL3 (2-D10, V_f= 1%)

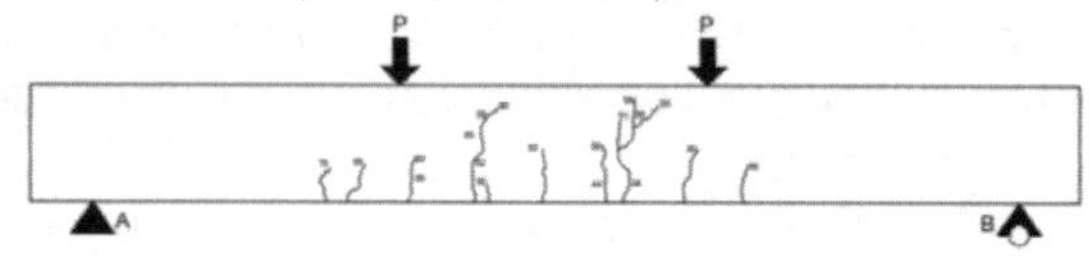

d. BL4 (2-D10, V_f= 1.5%)

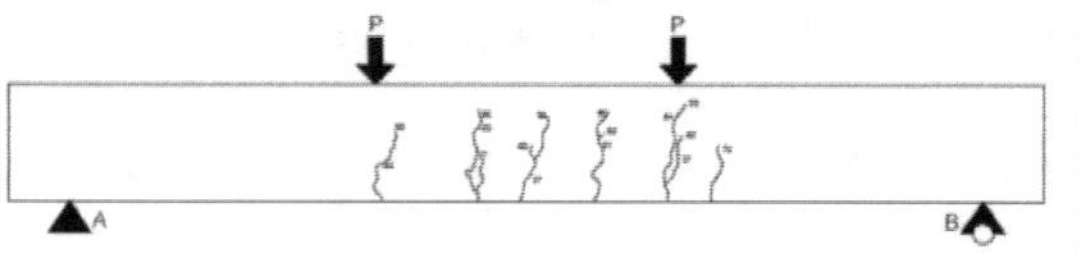

e. BL5 (3-D10, V_f= 1%)

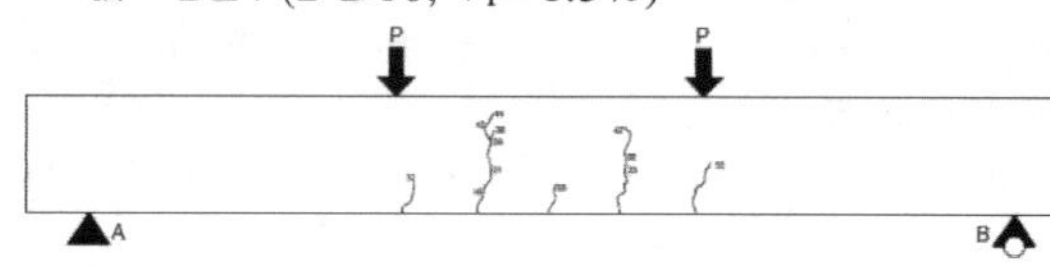

f. BL6 (2-D6, V_f= 1%)

Figure 5. The crack-pattern of the beam specimens

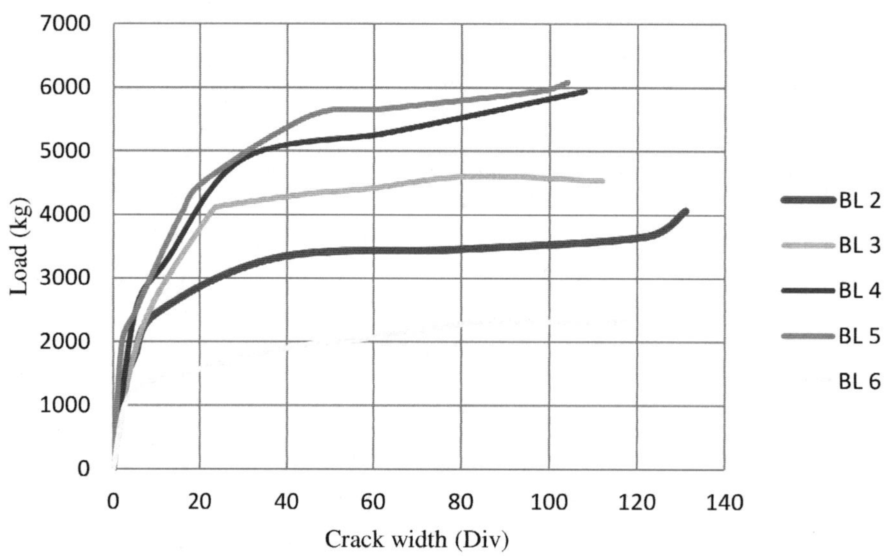

Figure

Figure 6. The load-crack width relationship of the beam specimens

The data of the maximum load of the reinforced concrete beam specimens as well as the percentage increase can be seen in Table 4. While, the load-deflection curve of each beam are presented in Figure 7.

From Figure 7 it can be seen that the addition of steel fibres to reinforced concrete beams can increase the beam stiffness and the maximum load capacity of the beam with the similar maximum deflection, as shown by the BL1 - BL4 curves.

The addition of V_f = 0.5% of steel fibres in the reinforced concrete beam specimen only has a very small effect on the stiffness and maximum load capacity of the beam, which is about 7% (see BL1 and BL2 curves in Figure 7). While, a significant increase in the maximum load occurred in BL4 specimens (V_f = 1.5%), which was 56% compared to BL1. However, as has been described above, at this fibre volume fraction, the workability of the fresh concrete is no longer acceptable.

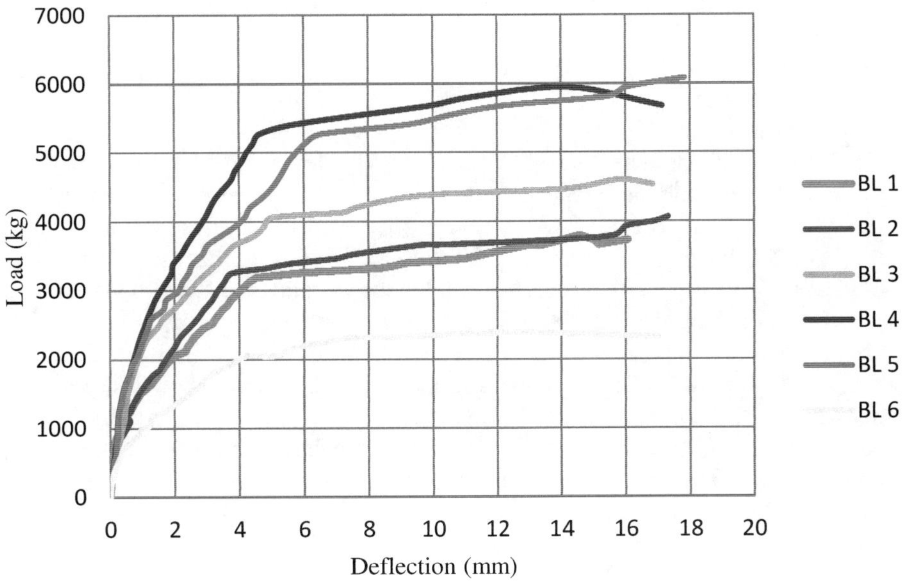

Figure 7. The load-deflection relationship of the beam specimens.

As has been described above, BL4 contained 2-D10 of flexural reinforcement and $V_f = 1.5\%$ and BL5 contained 3-D10 of flexural reinforcement and $V_f = 1\%$. It can be seen that both have almost the same maximum load. So it can be said that the addition of V_f from 1% to 1.5% can provide the same performance by adding one D10 of the flexural reinforcement (from 2D10 to 3D10). Or in other words in this case, the addition of $V_f = 0.5\%$ of steel fibre is equal to the addition of 1D10 of flexural reinforcement ($\rho = 0.0033$) to the beam.

BL6 curve (with 2-D6 of flexural reinforcement and $V_f = 1\%$) has lower stiffness and maximum load (-38.94%) than the reference specimen (BL1). The BL6 specimen has a flexural reinforcement ratio (ρ) that is smaller than the minimum reinforcement ratio (ρ_{min}). So it can be concluded that the existence of steel fibres could not improve the flexural performance of the beam that has $\rho < \rho_{min}$.

In general, the addition of steel fibre can increase the flexural load capacity and stiffness without reducing beam ductility.

4. Conclusions

The conclusion that can be drawn from this study are:

1. The addition of steel fibre with the volume fraction of 1% is a moderate value because it provides sufficient workability without the use of a superplasticizer and provides a fairly good strength.
2. The addition of steel fibres to the concrete mixture does not increase the compressive strength but increases the flexural tensile strength and change the concrete into a more ductile material.
3. The width of the cracks that occur is smaller, the first-crack load, the yield load and the ultimate load are getting higher along with the higher the steel fibre volume fraction contained in the beam.
4. The stiffness of the beam specimen increases with the increasing of steel fibre volume fraction without reducing its ductility.
5. Steel fibre contained in the beam with the flexural reinforcement ratio (ρ) less than ρ_{min} is not sufficient to improve the flexural performance of the beam.
6. In the case of BL4 and BL5 specimens, it was found that an increase in $V_f = 0.5\%$ from the previous results in a similar flexural performance with the addition of 1 bar flexural reinforced concrete beam.

Acknowledgments
This research was sponsored by the Faculty of Engineering, Universitas Lampung (DIPA FT 2020).

5. References

[1] Hameed R, Sellier A, Turatsinze A, and Duprat F 2013 Flexural behaviour of reinforced concrete beams: experiments and analytcal modellin. *Pak. J. Engg. & Appl. Sci.* Vol. 13 (p. 19-28).
[2] Altun F, Haktanir T, and Ari K 2007 Effect of steel fiber addition on mechanical properties concrete and RC beams. *Construction and Building Materials Journal* p. 654-661.
[3] Lim T Y, Paramasivam P, Lee S L 1997 Behavior of reinforced steel-fiber-concrete beams in flexure *J. Struct. Eng.* 113: 2439-2458
[4] Oh B H 1992 Flexural analysis of reinforced concrete beams containing steel fibers
[5] Nicolaides D and Markou G 2015 Modelling the flexural behaviour of fibre reinforced concrete beams with FEM *Engineering Structures* 9 653-665
[6] Hwang J H, Lee D H, and Kang J W 2013 Shear behavior models of steel fibre reinforced concrete beams modifying softened truss model appoaches *Materials Journal* 6 4847-4867.
[7] Naaman A 2007 Deflection-Softening and Deflection-Hardening FRC Composites: Characterization and Modeling. *Special Publication* 248: p. 53-66.

AUTHOR INDEX

Institute of Physics
Dirac House, Temple Back
Bristol BS1 6BE UK

ISSN: 1757-8981
ISBN 978-1-7138-4690-1